'Many of my critics considered that there was no future in tractor production and others thought it nothing short of madness... We had our difficulties, of course, particularly in those very early days, but the lusty infant not only survived, it flourished.'

SIR DAVID BROWN

The David Brown Tractor Story
Part One 1936–1948

STUART GIBBARD

Old Pond Publishing

First published 2003

Copyright © Stuart Gibbard 2003
The moral right of the author in this work has been asserted

All rights reserved. No parts of this publication may be reproduced, stored in a retrieval system, or transmitted, in any form or by any means, electronic, mechanical, photocopying, recording or otherwise, without prior permission of Old Pond Publishing.

ISBN 1 903366 39 9

A catalogue record for this book is available from the British Library

Published by

Old Pond Publishing,
104 Valley Road, Ipswich IP1 4PA United Kingdom

www.oldpond.com

Frontispiece illustration:
David Brown VAK1 tractors in the dispatch area of the tractor assembly shop at Meltham Mills during the Second World War.

Cover design and book layout by Liz Whatling
Printed and bound in Great Britain by
Butler & Tanner Ltd, Frome & London

Contents

Acknowledgements 6
Author's Note 7
Preface 8
Foreword by Adam Brown 9

Chapter One	Mastery in Gears	10
Chapter Two	Ferguson and Brown	32
Chapter Three	Swords before Ploughshares	58
Chapter Four	Tractors at War	94
Chapter Five	Power for Peace	148

Appendix 1	David Brown Tractor Serial Numbers 1936 – 1948	206
Appendix 2	VAK1 and VAK1/A Series Implements and Auxiliaries 1939 – 1947	208

Bibliography 209
Index 210

Acknowledgements

The preparation of this book has taken longer than any I have ever written. This was partly because the road taken by David Brown during the course of its tractor production involved many blind alleys, but also because I was overwhelmed by the number of former Meltham personnel who were both willing and keen to talk to me about their experiences. No way could I pass up such an opportunity to gain so much first-hand knowledge.

I have always had an affinity with Yorkshire (my father comes from Doncaster and my wife is a Skipton lass) so I already knew how warm and welcoming the people who hail from that county are. Even though many of the personnel who worked at Meltham were not Yorkshire born and bred, they have adopted its ways and my trips 'up north' became a most pleasurable experience.

The David Brown men made me feel at home and went to great lengths to ensure that I saw all the right people and got all the right information; their hospitality was genuine and their help was tireless. I was also struck by the great depth of their experience and knowledge. This, I can only put down to the fact that many of them worked for David Brown for most of their lives, from joining from school as an apprentice to leaving at, or more often beyond, retirement age – something that is almost unique today.

I would like to start by paying particular thanks to the 'senior' team of Charles Hull, Fred and Connie Meadowcroft, Steve Moorhouse, Arthur Caudwell and Len Craven, who have all gone out of their way to help and have been a tireless source of information. The following have also kindly given me a great deal of their time and provided a wealth of material and detail through their reminiscences: Derek Bennett, Derek Marshall, John Nicholls, Richard Garratt and John Dyson.

Sadly, I lost two of my valuable contacts during the preparation of this book when Herbert Ashfield and Commander Bill Ladbrooke passed away. Both had provided me with much useful information, which I was luckily able to build on with the kind assistance of Herbert's sons, Robert and Paul, and Bill's wife, Joyce.

Without turning this into a veritable 'Who's Who' of David Brown men, I would also like to thank Adrian Rubie, John Crabtree, Charles Clayton, George Franks, John Gant, Brian Edwards, Alf Wolton, Stewart Barnet, David White and Granville Wilson. While many of these last mentioned are too young (!) to feature in this book, they, and others, have all provided valuable material and contacts and their time will come in subsequent volumes. Keep the information coming, there are two more books to go!

I have reserved special thanks for Peter Murray and that indomitable double act of Bob Marsh and Eric Hirst. All three have been helpful to the extreme, searching out people and places, chasing up contacts and addresses and even acting as unofficial tour guides to the area.

Like many former David Brown personnel, Peter, Bob and Eric now play an active part in the David Brown Tractor Club. The club has given me tremendous support and I could not have written this book without its help. In particular, I would like to thank its chairman, Darrel Clegg, who has loaned me much valuable material from his own personal collection. Thanks also to Anthony Marsh, Neil Singleton and Mark Birkenshaw. Several generations of Darrell's family worked at the Meltham plant; the club is in good hands and I would highly recommend anyone who has even a passing interest in David Brown to join.

I would like to acknowledge the support of the David Brown Group at Park Works, now a Textron company, and thank John Hudson, whose encyclopaedic knowledge of David Brown, the man, the products and the company is second to none. Thanks also to Case International Ltd and Steve Crowe and Phil Brown at Basildon.

This book is very much a record of one man's goal to build the best tractors in the world. David Brown Tractors Ltd was very much Sir David's 'baby' and I could not have told the story without an understanding of his character and motivation. That insight was provided by his grandson, Adam Brown, who also very generously allowed me access to family records and kindly agreed to write the Foreword.

James Cochrane and Norman Cundy, two true enthusiasts, must collectively take the blame for insisting that I turn my attention to David Brown. Having said that,

I could never have attempted to write the book without their dedicated assistance, for which I am most grateful. Patrick Palmer has very kindly allowed me to borrow a wealth of material from his collection while supplying useful background information on his brother, Nigel. I would also like to thank Selwyn Houghton for once again sharing his extensive knowledge of Ferguson-Brown tractors.

To adequately cover the full scope of David Brown's extensive and varied product line requires specialist knowledge and that was forthcoming from David Fletcher at the Tank Museum, David Morris at the Fleet Air Arm Museum, Christine Gregory and Matthew Barrows at the Royal Air Force Museum, Lynne Thurston at the National Railway Museum and staff at the Royal Engineers Museum. Fellow author Stuart Broatch has also briefed me on the Vauxhall and Bedford story.

I am grateful for the assistance of Jonathan Brown and Caroline Gould of the Rural History Centre at the University of Reading for help in accessing their extensive David Brown records. Special thanks to my colleague, Tony Clare, for a very useful exchange of information on Dr Merritt and the Nuffield/David Brown connection. I must also thank my old friend, Peter Love, the editor of *Tractor & Machinery,* for everything from running errands to supplying and identifying photographs.

So many people have helped in the preparation of this book and thanks are also due to Norman Aish, James Baldwin, Nick Baldwin, Ven Dodge, Dr Hugh Kearsey, Sam Kennedy, Alan Kersey, John Moffitt, Jim Russell, John Sayles (McCormick Tractors International), Ben Serjeant, Graham Scarboro (Maltby Group), Paul Wade (McCormick Tractors International), Charles Weight (Tractor Spares Ltd) and Dave Woolerton. I hope I have not left anyone out. If I have, then I apologise for it was not intentional.

Finally, the usual thanks to the highly efficient production team including Roger Smith and Felicity Halston at Old Pond, Julanne Arnold, Liz Whatling and my wife, Sue, who has prepared the index and has actually had to read one of my books for the first time!

STUART GIBBARD
June 2003

Author's Note

In our present age of enforced metrification, it is satisfying to discover that there was just as much confusion between imperial and metric measures seventy years ago as there is today. Most British manufacturers gave their vehicle specifications in imperial figures but, for some unknown reason, turned to metric when it came to engine capacities, which were traditionally given in litres or cubic centimetres. Vauxhall was almost unique in describing the capacity of its vehicle engines using the imperial units of cubic inches, as was the preferred method in the USA. For a time, David Brown seems to have followed Luton's lead and for that reason I have remained faithful to imperial units but have also given metric engine capacities in parentheses.

Horsepower figures, where possible, are taken from the manufacturer's literature. All are approximate. Several different systems for calculating engine power have been used over the years and only those figures I can verify as true brake horsepower (bhp) are given as such.

One of the leading personalities in my narrative is Sir David Brown, the late chairman and managing director of the David Brown Corporation. As he was not knighted until 1968, he appears for the period of this book as plain *David Brown* without the honorific. In an attempt to avoid confusion between David Brown the man, David Brown the company, and David Brown the tractor, I have usually referred to him as *Mr* David Brown, as he was respectfully known to many of his staff before his knighthood.

Preface

I have always been fascinated by David Brown and have owned numerous models from a VAK1 to a 1390. As a tractor company, it seemed such an enigmatic concern, far removed from the brash image of the mainstream manufacturers with their big-city factories and profit-led philosophies. Its plant was tucked away in a remote Pennine valley and the firm seemed to exude that great British tradition of putting engineering excellence before monetary considerations. Here was no faceless conglomerate but a proud institution run by people who really cared about the company, the product and the customer, and actually believed that they were building the best tractors in the world.

Because of its low-key image, the marque has not always been given the prominence it deserved. For many years, David Brown was the 'Cinderella' of the vintage tractor movement, partly because of the belief that the hunting pink tractors were too modern to merit preservation. In many ways, the make became a victim of its own achievements.

I remember when I was hit by the tractor-collecting bug in the early 1970s, I went out and bought three tractors in as many weeks. They were an Allis-Chalmers Model B, a Fordson E27N Major and a David Brown Cropmaster. All were obtained at local farm sales and none cost me more than £50 each. When I arrived home with the David Brown, our farm foreman suggested that it was 'too new' to bother with and never really believed me when I told him that it was actually older than either the Allis or the Fordson. He never said any more once he realised that the Cropmaster made a useful spare tractor on the farm.

Since then, everything has changed and today the David Brown tractor is appreciated for what it is – a machine with a rich heritage of British engineering innovation at its best. It was always ahead of its time in terms of both styling and specification and the company remained at the vanguard of technical development.

The list of 'firsts' pioneered by David Brown is endless, from the turnbuckle top-link to the use of dished wheel-centres to alter track width – all simple features that we take for granted today. It could also be argued that if Sir David Brown had not had the confidence and foresight to press ahead with Ferguson's designs in the 1930s, then the three-point linkage or draft control might never have seen the light of day.

Part of David Brown's success was undoubtedly due to its location. The community that had once been dominated by the cotton industry threw itself unreservedly into tractor production. Meltham was under no illusion that its prosperity rested on the success of the hunting pink machines and the townsfolk became accustomed to seeing lines of red (and later white) tractors trundle past their windows. It was a rural area and many of the factory workers had allotments or smallholdings and understood all about tractors and how they worked, and that knowledge undoubtedly helped sustain the level of craftsmanship for which the factory was recognised. Meltham became David Brown and David Brown *was* Meltham.

This book is the first of three volumes. Why three books, you may ask, when I have devoted less to other leading (and better known?) manufacturers in previous works? Well, I thought I knew a little about David Brown until I began researching this book. I then quickly realised that my limited knowledge ran only as far as the tip of the iceberg and that the scope of the company's exploits was truly immense. One book became two and then three as I realised that this was the only way I could do full justice to the David Brown story.

Taking the year 1956 as an example, Ford and Ferguson had just one tractor model (with minor engine variations) each in production in the UK. David Brown, on the other hand, was marketing an almost unbelievable range of wheeled and tracked agricultural tractors, industrial crawlers and equipment, aircraft towing tractors, farm implements and accessories, stationary and marine engines and even a specialist market-garden tractor.

I also found it impossible to divorce tractor production from everything else that was going on within the David Brown organisation. The history and affairs of the parent group, gear manufacturing, even Aston Martin, all had a bearing on what happened at Meltham. Neither could I ignore the man at the helm; David Brown

Tractors Ltd owed its success to the personal presence of Sir David Brown and, for most of its existence, it operated under his guidance. I therefore make no apology for including and documenting within the book the background to David Brown, the man, the companies and the machines, because I strongly feel that the three formed an interconnected and unbreakable trinity at the heart of the tractor story.

This first book chronicles the history of the company from its earliest days, up to the introduction of the first true David Brown tractor, the VAK1 model, and the subsequent first ten years of production at Meltham. The Second World War saw the fledgling tractor division thrown in at the deep end and almost immediately having to deal with massive defence contracts and extended production schedules. The company coped admirably and emerged to become one of the major players of the post-war British tractor industry. These formative years were one of the most important periods in the history of David Brown Tractors Ltd and I have tried to document the full extent of its wartime production to give some indication of the type of pressures the company was under.

As a final point, I would like to suggest that if you have never made the pilgrimage to Meltham, then you should go and visit the factory site and look at the remaining buildings; compare the size of the plant with the size of the town and you will begin to understand what David Brown meant to the community. Stand in the bar at Durker Roods (now a hotel) and listen: before long the words 'David Brown' or 'tractor' will be heard in the general buzz of conversation, even though it is twenty-five years since the plant closed. Talk to people who worked at the factory and you will find it difficult not to be swept up by their dedication to the old tractor company, or moved by their lasting affection for the works, the assembly line and the machine shops. When you come away, you too will believe that David Brown probably did make the best tractors in the world.

STUART GIBBARD
June 2003

Foreword

As Honorary president of the David Brown Tractor Club, I can think of many enthusiasts who will enjoy and treasure this book on David Brown. It will appeal to a much wider audience too. David Brown created a worldwide business whose byword was quality whether the product was a gearbox, an Aston Martin car or a David Brown tractor.

This book takes us back to an age when engineering drawings were produced with pen and ink; when mathematical calculations were made on the slide-rule. The characters in the book were the pinnacle of the mechanical age; all were top-class engineers and the product reflected that inherent quality.

Small, dapper and shunning publicity, Grandfather had the steely determination and clarity of thought that drove him to achieve competence at many accomplishments. He could hunt to hounds (he was Master of the Badsworth and South Oxfordshire), play polo, play tennis and fly a plane. Above all, he created a business that became a legend; yet his motto was typically unassuming: 'One's chief aim in life is to make two blades of grass grow where one grew before.'

ADAM BROWN
Upper Denby, Huddersfield

CHAPTER 1 — *Mastery in Gears*

The heavy machine shop at Park Works. By 1920, David Brown & Sons Ltd was one of the world's largest gear manufacturers.

RIGHT
David Brown, the founder of the family firm, was a skilled pattern maker who set up business in Huddersfield in 1860.

The David Brown organisation was born not out of iron or steel but from wood. The founder and the man from whom the company took its name was a skilled joiner who specialised in pattern making – crafting the wooden patterns that were used to form the moulds for casting iron.

David Brown, whose father, John Brown, was also a joiner, was born on 17 February 1843 at New Bank, Northowram, near Halifax. In 1860, when he was only seventeen years old, he set up shop in Vulcan Road, Huddersfield, above old stables belonging to a local ironmaster, Thomas Broadbent. Trading as Brown & Broadbent, David made the gear patterns for Broadbent's foundry, which supplied gears for looms and weaving machinery.

As the business prospered, David Brown expanded into larger premises on the site of Broadbent's foundry in Chapel Street. On 31 October 1867, he married a local girl, Mary Jane Matthews, who came from Shepley, near Holmfirth. They both moved in with Mary's parents to live at South Street, Huddersfield, where their first son, Ernest, was born in 1871.

The mid-nineteenth century saw Britain in the grip of the Industrial Revolution. Huddersfield was at the core of the industrial heartland of central Yorkshire – an area whose prosperity was built on wool. Woollen and worsted mills were all around and growing in number. Far from being William Blake's 'dark Satanic mills', these were nonetheless imposing structures of local stone, nestling comfortably against the stark backdrop of the Pennine hills.

Wool and cotton were becoming increasingly important industries. By the middle of the nineteenth century there were nearly as many textile workers in the country as there were people employed in agriculture, and the worsted mills in Yorkshire were importing raw materials from as far afield as Australia and South America.

This economic expansion, which began in the eighteenth century, saw Huddersfield grow from a large village into an important town with a thriving textile industry. The local economy was strengthened by the construction of the Calder and Marsden canals and good rail links, which encouraged the establishment of further ancillary and heavy engineering industries. Against this background, there were great opportunities for those astute enough to take advantage of the industrial development. David Brown was such a man, and by 1873, the year his second son, Frank, was born, he had expanded into gear manufacture – supplying all kinds of spur, skew, bevel and eccentric gears. He manufactured not only wooden cogs but also cast-iron gears, which were made for him by Broadbent and other local foundries.

In 1879, Brown established his firm as David Brown & Company with the office registered at his home address of 72 South Street. Increased mechanisation of the textile mills meant that iron gears (stronger and more regular in motion than their wooden counterparts) and the patterns to make them were in great demand. By 1890, David Brown was employing ten men, including his two eldest sons, Ernest and Frank. Following a fire at the Chapel Street works in 1895, the company fell out with Broadbent and moved to new premises in East Parade. Frank Brown took over the responsibility for the company finances, leaving Ernest to look after the heavy patterns, while

LEFT:
David Brown & Sons' East Parade Works in 1902, seven years after the firm moved into the premises. East Parade continued to operate as a pattern works under the direction of Ernest Brown after the Lockwood factory was opened in 1903.

David Brown's youngest son, Percy, joined the business to handle the wooden patterns. After their father retired in 1898, the three brothers formed the business into a partnership known as David Brown & Sons

Machine-cut Gears

After the move to East Parade, Frank and Percy Brown were eager to expand and felt that the company's future lay in machine-cut gears. A greater understanding of the chemistry of iron manufacture at the end of the nineteenth century, together with advances in the fields of electrification and internal combustion engines during a period often dubbed the scientific revolution, was leading to fundamental changes in engineering practices. The improvisation of the early industrial pioneers was giving way to scientific calculations and theory. Greater accuracy meant finer tolerances and precision engineering demanded precision gearing.

Following experiments with a machine built by the company itself to cut teeth in both wooden and metal gear patterns, Frank and Percy decided to invest in the necessary manufacturing equipment for machine-cut gears, placing orders for turning, cutting and milling tools with firms in the USA. The machinery was installed in the basement at East Parade and the company began cutting its first gears on 2 January 1899. After a shaky start, the new venture proved to be a very profitable sideline, bringing the firm both more work and greater recognition. Within two years, the original workforce of six men employed on the gear-cutting side had risen to over thirty.

David Brown died in 1901, aged only sixty years, having never really recovered from the shock of an earlier accident with a machine tool that had severely maimed his hand. After his death, the three sons remained in partnership but divided the responsibility for the company between them. Ernest carried on with the gear patterns at East Parade, leaving his two brothers to pursue the gear-cutting business.

Frank and Percy's immediate concern was to find suitable premises in which to expand their manufacturing facilities. After some searching, they found a greenfield site at Lockwood on the

BELOW:
Park Works at Lockwood, showing the machine shops and the Park Road offices at the time of the factory's opening in 1903. Park Cottage can just be seen in the background to the left of the photograph.

RIGHT:
David Brown & Sons issued its first catalogue for machine-cut gears in 1903.

BELOW:
Milling the threads on David Brown worm-shafts. Park Works pioneered new gear-cutting techniques and developed its own manufacturing machinery.

south-eastern outskirts of Huddersfield, just two miles out of the town. The site, a 15 acre parcel of open heathland together with Park Cottage, a beautiful old house with a long drive and a lawn surrounded by iron railings, was purchased from the Whitely family in 1902. The first works buildings, including a ground-floor office block with a basement and four bays of machine shops, were erected in the bottom corner of the field. The new factory was christened Park Works and Frank Brown later moved into the cottage to oversee the new facilities.

In 1903, the company was incorporated as David Brown & Sons (Huddersfield) Ltd. Park Works began manufacturing gears and issued its first general gear catalogue. Over the next few years, Frank and Percy pioneered new types of worm and helical gears, introduced new gear-tooth patterns and explored ways of improving gearing efficiency. This led to increased business and marked the beginning of a period of major expansion for the company.

The manufacturing capacity at Park Works was increased in 1910 with more machine shops

LEFT:
Expansion at Park Works saw a new iron foundry established on the site in 1912. The workers are seen preparing the moulds for casting.

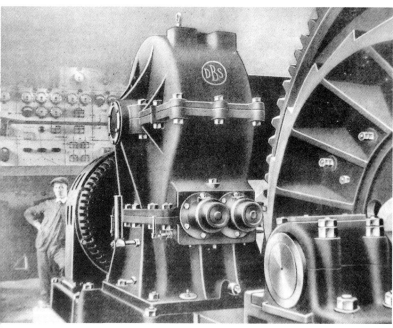

LEFT:
By 1912, David Brown & Sons was the largest gear manufacturer in the British Commonwealth. Its products included this duplex worm reduction unit used to drive a 150 hp endless-rope haulage system.

erected. The shops were heated by either low-pressure steam or hot water and were lit by electric lighting. Electricity obtained from the Huddersfield Corporation also powered all the works machinery. An iron foundry was established on the site two years later.

David Brown & Sons was now the largest gear-manufacturing concern in the British Commonwealth. By 1912, its turnover had reached £143,000, which was a considerable achievement for the time. The same year, Frank and Percy decided to turn the gear firm into a public company to secure capital for future expansion. Frank Brown also made several trips overseas to study both European and American methods of manufacture. In 1913, he concluded an agreement with the Timken Detroit Axle Company of Canton, Ohio, to manufacture worm-gear drives for rear axles in the USA.

The David Brown Patent Thread

Much of David Brown's early success was down to its development of a new type of worm-gear thread. Worm gearing in its earliest form was a direct development of the simple screw and could be traced back to the work of Archimedes in around 200 BC. It became a relatively simple method of transmitting power from one shaft to another when both were not in the same plane.

However, before the advent of accurate gear production methods, worm gearing was a very inefficient mechanism. The straight-sided worm threads caused friction, leading to a loss of power, overheating, undue wear and noisy running. David Brown's engineers, led by the works manager, F. J. Bostock, discovered that by altering the shape of the worm threads by evolving an entirely new thread profile with an involute curve, they could convert the rubbing action of the worm into a rolling movement and greatly improve its efficiency.

RIGHT
David Brown & Sons introduced an involute helicoidal thread for worm gearing in 1914 and patented it the following year. The design was more efficient than the earlier worm gears and was almost silent when running.

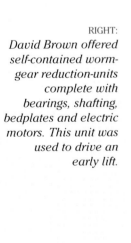

RIGHT:
David Brown offered self-contained worm-gear reduction-units complete with bearings, shafting, bedplates and electric motors. This unit was used to drive an early lift.

David Brown & Sons' research into worm gearing eventually led to the introduction of the involute helicoidal thread in 1914. Patented the following year, this revolutionary and highly efficient design gave nearly silent running and was much easier to lubricate, thus reducing wear to a minimum. The thread was tested at the Daimler-Lanchester plant at Teddington and proved to be 97.4 percent efficient. It became known as the David Brown Patent Thread and was eventually adopted as the British Standard thread.

The worms were cut on a thread-milling machine while the tooth profile and shape on the gear wheel were obtained by using a special form cutter, known as a hob, which was basically a replica of the worm with cutting edges. Because it was at the forefront of modern gear design, David Brown & Sons had to develop its own manufacturing machinery, which was later patented and supplied to other manufacturers.

The company also introduced self-contained worm gear and reducing gear units complete with bearings and shafting. Bedplates and electric motors were also supplied if required. The applications were endless and the timing could not have been more opportune with most factories moving over to electrification. There was also a growing demand for worm-drive axles for motor vehicles.

David Brown had patented its first form of rear-axle worm gears in 1913, and within a short time these units had become standard on all buses operated by the London General Omnibus Company. Motor buses had struggled to gain acceptance in London because of the noise they created, but the famous B-type 'London General', built for LGOC by the Associated Equipment Company of Walthamstow using the David Brown worm-drive, was praised for the silent running of the steel worm in its rear axle. AEC's bus was quieter than most of its competitors and was adopted by most of the major omnibus fleet operators in the capital. The coming of the electric trolley-bus led to further orders for Park Works because the system

ABOVE:
The Associated Equipment Company standardised on David Brown worm-drive rear axles for its early lorries and buses. This AEC Y-type truck dates from around 1919.

pioneered in nearby Keighley and the Lancashire town of Ramsbottom, also relied heavily on the David Brown worm-drive axle.

David Brown Cars

One of David Brown & Sons' first ventures outside gear manufacture was a short-lived incursion into motor car production. In 1908, having already abandoned a couple of prototype vehicle designs of its own, the company entered into an agreement with a Cambridgeshire engineer, Ralph Lucas, to manufacture his Lucas Valveless car at Park Works.

Lucas had been experimenting with cars since 1901. His first design employed a strange two-stroke engine with an opposed piston and crankshaft at either end of a single cylinder. It would run on paraffin and the vehicle was reversed by bringing the engine to an idle and then advancing the ignition, causing it to backfire and run backwards. He later played around with several types of transmission using various complicated dog-clutch, chain and sprockets or bevel gear arrangements, and began sketching out a new engine layout.

After Lucas joined forces with David Brown, the Huddersfield company employed a resourceful engineer by the name of Frederick Burgess to turn Lucas's rough sketches into reality. Burgess refined the two-stroke engine into an unorthodox transverse unit

LEFT:
A taxicab version of the Valveless car. The Valveless, which was built at Park Works from 1908 to 1915, was renowned for its simplicity, reliability, low fuel consumption and silent running.

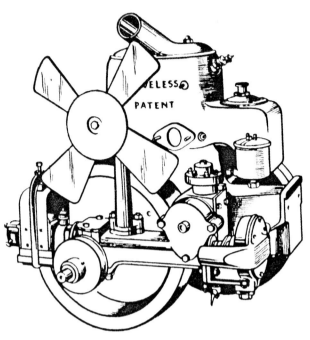

RIGHT:
Designed by Ralph Lucas and improved by Frederick Burgess, the Valveless two-stroke power unit was an unorthodox transverse engine with two pistons sharing the same combustion chamber. It had two contra-rotating crankshafts and flywheels.

with two vertical pistons that reciprocated simultaneously. They shared the same combustion chamber and there was only one spark plug. The two cast-iron pistons each had their own crankshafts, which revolved in opposite directions in an aluminium crankcase, and the cylinders were provided with inlet and exhaust ports but no valves, hence the name 'Valveless'.

Lucas's ambition had been to design an engine of extreme simplicity. It had only six working parts: two pistons, two connecting rods and two crankshafts. The crankshafts were fitted with heavy flywheels, which were geared together and weighed around ¾ cwt each. The effect of the two contra-rotating flywheels was to reduce engine vibration, ensuring smooth, almost silent, running with remarkable hill-climbing abilities.

Two models of the Valveless were made at Park Works: a 19.9 hp version with a 5 in. bore and a 5 in. stroke, and a 15.8 hp car with a slightly reduced bore. Both had a conventional leather-to-metal type clutch, a four-speed and reverse gearbox and a David Brown worm-drive rear axle.

Valveless Cars Ltd of Huddersfield was set up as a David Brown subsidiary with showrooms, run by a Mr. Dodson, at Bond Street in London. The car was praised for its simplicity and reliability, and one example evidently covered over 70,000 miles without any problem. Fuel consumption was a creditable 20 miles per gallon, and the only criticisms of the Valveless were its lack of acceleration and its tendency to backfire when

RIGHT:
A Dodson car outside Park Works in about 1912. This British copy of the French Renault was built under licence by David Brown from 1910 to 1914. Note the gear wheels set into the factory gates.

LEFT:
Frank Brown's wife, Carrie, at the wheel of her Valveless tourer with their son, David, in the passenger seat. Mrs. Brown was the second woman in Huddersfield to obtain a driver's licence.

pushed hard. Frederick Burgess tested every chassis that came out of Park Works, using an upturned orange box as a makeshift seat, and personally delivered some of the cars to the London showroom. Around 800 are believed to have been built before production came to an end in 1915.

Park Works also supplied worm-drive rear axles for the 18 hp Sava motor car, manufactured from 1910 by the Société Anversoise pour Fabrication des Voitures Automobiles of Antwerp. The David Brown company was persuaded to take an agency for this obscure Belgian marque and sell it alongside the Valveless. However, it created little interest, and the Sava agency was dropped almost immediately after the manager of David Brown's Bond Street showrooms, Mr. Dodson, managed to secure the rights to build the French Renault car under licence.

This British copy of the Renault, almost identical to the French car except for the provision of a Zenith carburettor, was put into production at Park Works in 1910. It was marketed as the Dodson and two models were available, the 12/16 hp with a three-speed gearbox and the four-speed 20/30 hp.

The Dodson's antiquated quadrant gear-change made it unpopular and few were sold. It was dropped in 1914 after the outbreak of the First World War brought an end to David Brown car production, although several Dodson chassis still remaining at Park Works were later converted into military ambulances.

Frank Brown never learnt to drive but employed a chauffeur to ferry him about in a Dodson limousine. His wife, Carrie, on the other hand, became only the second woman driver to get a licence in Huddersfield (and possibly Yorkshire) and ran a Valveless car for many years. No examples of the Dodson vehicles have survived, but three Valveless cars still exist as evidence of David Brown's brief and ultimately unprofitable foray into early motor car production.

When car production at Park Works was wound up, Frederick Burgess joined Humber in Coventry as a designer and works driver and was responsible for the twin overhead-cam engine that was fitted to Humber's 1914 TT (Tourist Trophy) racing cars. While at Humber, he met Walter Owen Bentley, a former Great Northern Railway apprentice who had modified French Clerget rotary aero-engines during the First World War. In January 1919, after leaving Humber, Burgess and

Bentley joined forces to begin working on a new three-litre car engine, which ran for the first time nine months later. The following year, W. O. Bentley formed Bentley Motors Ltd of Cricklewood, London, with Burgess as his chief engineer.

Growth and Consolidation

The First World War saw David Brown & Sons engaged in a variety of work for the war effort. Having developed a range of high-speed turbine gears, the company was asked to manufacture a series of marine propulsion units for all classes of warships from destroyers to motor-patrol boats and even submarines. The propulsion units were supplied to most of the major shipyards, notably the Wallsend Slipway, Richardson Westgarth, John Samuel White, J. I. Thorneycroft, Yarrow and the Fairfield Shipbuilding Company.

Other important developments to come out of Park Works included a shell-boring lathe and gun training and elevating mechanisms. The company also supplied paravanes, Stokes trench-mortar bombs, funnel tilting gear and helical-geared bollards for warships, as well as the axles and final-drives for Thornycroft army lorries. This extensive war work saw the number of employees at Park Works increase from 200 to over 1,000.

Orders continued to flow in after the war, leading to the need for further expansion at Park Works. A bronze foundry, heavy-machine shops, pattern shops and stores including some of the earliest

RIGHT: *During the First World War, David Brown & Sons developed a series of marine propulsion units for all classes of warships including the Royal Navy's latest submarines.*

RIGHT: *Important war production carried out by David Brown & Sons at Park Works during the First World War included the manufacture of rear axles and final-drive units for Thornycroft army lorries, such as this J-type model.*

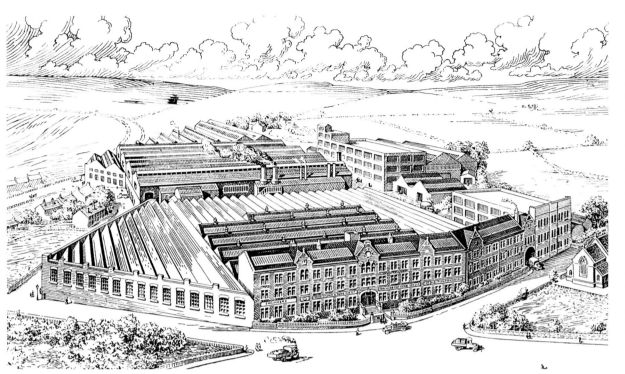

LEFT:
Dating from around 1920, this artist's impression of Park Works shows how the plant had expanded in the immediate post-war period. Note the extra two stories added to the offices. The three-story concrete building (on the right towards the rear of the site) was a pattern store for the iron foundry, which can be seen in the centre of the works.

BELOW:
The main office at Park Works. The administration block had been extended during the 1920s by the addition of two extra stories over the original office building.

pre-stressed concrete buildings in the country, had all been added by 1920, with each expansion having to overcome the complexities of building on a sloping site. The offices were also extended into a new administration block by the addition of two extra stories and were sumptuously finished in oak panelling from Waring & Gillow of London, who also supplied the boardroom furniture.

By the 1920s, David Brown & Sons was

ABOVE:
Worm-gear assembly in the heavy fitting shop at Park Works. By the 1920s, David Brown & Sons had become the world's largest worm-gear manufacturer.

RIGHT:
A plan of the Park Works site dating from around 1929. The buildings covered over 12 acres and the plant employed nearly 1,500 men.

recognised as the world's largest worm-gear manufacturer. It was also supplying machine tools and had developed further new tooth forms, including double- and triple-helical gears. In 1921,

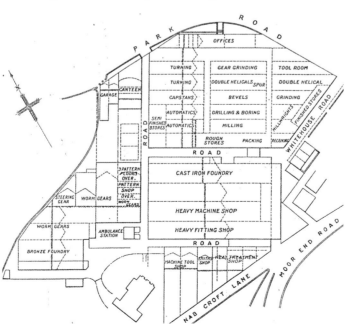

Frank's seventeen-year-old son, also called David, joined the company. The young David Brown was evidently a dynamic personality who had his own ideas for developing the business and was keen to move the firm in new directions. After successfully troubleshooting an ailing British subsidiary, the Keighley Gear Company, as well as sorting out problems with interests in France, he was invited to join the board of the parent company in 1929.

David Brown & Sons' turnover for that year was an impressive $530,000; the buildings at Park Works now covered over 12 acres and nearly 1,500 men were employed on the site. As the business grew, the company took over one of its competitors, the gear manufacturers F. R. Jackson Ltd, in 1930 and acquired a steel foundry at Salford in Lancashire in the process. Sadly, David Brown & Sons was to lose its chairman when Percy Brown passed away the following year

Frank took over the chairmanship, and his son, David, was appointed managing director in 1932.

David Brown 'the younger', as he was known by some of the older hands at Park Works, found that his first task was to consolidate the company in the face of the economic crisis that was gripping the country during the early 1930s. The slump was particularly affecting the core areas of David Brown's business, notably the iron and steel, shipbuilding and textiles industries, and the company's turnover for 1931 had slipped to £400,000. It was time to cut out the dead wood; overheads were trimmed and the workforce was sympathetically reduced with David Brown & Sons becoming one of the first firms to offer redundancy payments.

The new managing director also identified savings that could be made if the company became less reliant on outside suppliers for its castings. Greater foundry capacity was needed, but the Park Works site was already overcrowded. Thus, in 1934, David Brown purchased an old Cammell Laird plant at Penistone near Sheffield as the site for a new bronze and steel foundry. The plant came with twenty-nine acres of land and was bought for £17,500.

LEFT:

The Keighley Gear Company, an ailing subsidiary of David Brown & Sons, was successfully turned around by young David Brown between 1927 and 1928 to become a profitable venture producing medium-size gears.

BELOW:

The heavy-gear manufacturers, P. R. Jackson Ltd., together with its Salford works and foundry, were taken over by the David Brown group in 1930.

RIGHT:
The Penistone Works, David Brown's new steel and bronze foundry, opened in 1935. The 29 acre site was near Sheffield.

The Penistone site was very run down and there was massive unemployment in the area at the time of the acquisition. A report referred to it as 'a derelict heap of scrap in one of those forgotten villages that had died during the slump.' A contract for the demolition of the 110,000 sq ft main building had already been awarded to Thomas W. Ward Ltd, but David Brown's intervention won it a timely reprieve.

The new foundry opened in 1935, utilising one of the latest casting techniques that was pioneered in France and known as the Randupson process. This greatly simplified moulding process allowed David Brown to make use of the abundance of largely unskilled labour in the area. The derelict plant was transformed into one of the finest steel foundries in the world, converting Penistone into a prosperous township in the process.

RIGHT:
The Penistone foundry relied on a revolutionary new casting technique known as the Randupson process.

Once the Penistone operation was underway, Mr David Brown was able to close down the iron foundry at Park Works and reorganise the factory with a more co-ordinated layout, proper gangways and central stores. He also established separate product divisions and improved the efficiency of the sales and finance departments.

Men at the Top

There can be little doubt that it was Frank Brown's dynamic force that led to David Brown & Sons becoming established as one of the world's leading gear manufacturers. Born on 13 October 1873, Francis Edwin Brown, better known as Frank, was the second son of David Brown, the founder. He joined his father as an apprentice in 1889 and soon took over the responsibility for the financial side of the business.

Frank Brown married Carrie Brook, a local lass from Milnsbridge, on 29 June 1901. They lived at 72 South Street until after their son, David, was born on 10 May 1904. The family then moved into Park Cottage, where they remained until Frank bought a house in Holmfirth in 1921.

As joint managing director, Frank shared the running of the company with his younger brother, Percy, taking over the chairmanship after the latter passed away in 1931. His reputation as a tough businessman was justly deserved, but his brash demeanour and pugnacious appearance were tempered by his even-handedness; he showed great concern for the wellbeing of his workforce, who both revered and feared him in equal measures. It was said that he had a 'firm yet kindly hand'.

Leonard Craven, who worked for the firm for over forty years, provides the most vivid description of Frank Brown in his unofficial company history, *A Season of Growth:*

'Physically he was by no means a giant, but his stature was enormous. At least so it appeared to many of the rank and file who were left in no doubt that he was their employer... his frequent excursions through the works were to be likened in many respects to the passage of an eastern potentate through the bazaar. Fortunately for many, the grapevine was more efficient than the jungle drums, and his approach was invariably preceded by frantic preparations and precautions whilst his departure always left a rich aroma of Churchillian cigar smoke and sighs of relief from those who had survived the ordeal unscathed.'.

ABOVE:
Frank Brown in his office at Park Works. He became chairman of David Brown & Sons in 1931 and for many years was the driving force behind the company.

Although he liked to be kept abreast of agricultural and horticultural matters, Frank Brown had few interests outside the company and had no time for public life or hobbies other than the occasional drink and a game of bridge. Sunday mornings were spent walking around the factory (often with his young son, David, in tow) and his greatest ambition was to see David Brown & Sons' annual turnover exceed £1 million. He finally realised this in 1937.

Frank had a good personal and working relationship with his son, and supported most of young David's ambitious plans (publicly if not always privately) after the latter was appointed managing director in 1932. However, it has to be said that the relationship was sometimes volatile and often punctuated by massive rows.

Young David Brown entered the family business as an engineering apprentice in 1921 following a public school education at Rossall on the Lancashire coast. After gaining experience successively in the estimating department, drawing office, pattern shop and iron foundry, he travelled extensively in the USA, South Africa and Europe, studying factory methods and conditions.

Like his father, David Brown showed great

RIGHT:
Frank Brown's son, David, was fascinated with racing cars and built this trials special, based on Singer components, while working as an apprentice at Park Works in 1922.

determination and tenacity, but his (almost ruthless) ambition was often masked by his more relaxed appearance. He also liked to play as hard as he worked and had a passion for speed, kindled, perhaps, by early trips out with Frederick Burgess while testing the Valveless cars.

In 1921, David persuaded his father to buy him his first motorcycle, an 1100cc Reading Standard, which he used for hill climbs and speed trials. The following year, he designed and built his own racing car (much to his father's frustration) while working as an apprentice for the family firm. The car, a trials special based on Singer components, was assembled at Park Works and had a French Sage engine, a Meadows gearbox and a Timken rear-axle.

Even marriage to his childhood sweetheart, Daisy Firth, in 1926 failed to curb David's spirited behaviour. During that same year, Park Works received an order from Charles Amherst Villiers, a leading expert on forced induction, to build Roots-type superchargers for a Vauxhall TT racing car that was to be used in the Shelsley Walsh hill-climbs by the great Raymond Mays of ERA fame. This rare beast, originally built for the 1922 TT races, had a sixteen-valve, twin-overhead cam engine designed by Harry Ricardo. With the two Amherst Villiers

RIGHT:
Young David Brown successfully raced this supercharged Villiers Vauxhall racing car in northern speed trials and hill-climb events between 1926 and 1929. Developed by Amherst Villiers, it had a 280 bhp engine and a top speed of over 120 mph.

superchargers fitted, it developed 280 bhp and was capable of speeds in excess of 120 mph.

Frank Brown put David in charge of the project. When it came to testing, Amherst Villiers asked if there were any suitable venues in the Huddersfield area. David suggested the twisty Holme Moss road running through to Tintwistle between Holmfirth and Manchester. It was a road he regularly travelled with his motorcycle. When Raymond Mays failed to turn up for the first day's testing, Brown was allowed to take the wheel and set a time that even Mays could not better the following day.

The project was leading up to the development of the famous 4½ litre 'Blower Bentleys' that eventually raced at Le Mans, and Park Works was awarded the manufacturing contract for the superchargers for these prestigious vehicles. In recognition of young David Brown's input, Amherst Villiers agreed to sell him a second, almost identical, Vauxhall TT car provided that it was fitted with the same superchargers and carried the Villiers name on the bonnet. Brown used it very successfully in various northern speed trials and hill-climb events.

There can be no doubt that David's motor racing exploits were at odds with his business responsibilities and family commitments. He now had a son, also named David, who had been born on 23 December 1927 (his daughter, Angela, was born on 29 July 1932). After Frank Brown suffered a mild heart attack in 1929, David retired from racing to concentrate on the gear company. His father had evidently called him in and said, 'My son, if you are to be managing director of this business, you must give up motor racing'.

Twenty-eight years is a remarkably young age to be appointed managing director of such a large concern as David Brown & Sons, but young David Brown carried the mantle with considerable prowess. He had many outstanding qualities and was already a very shrewd businessman with great foresight and vision that even his father could not match. He was very focussed, almost driven, and never lost the desire to win, and it was this probably above all else that was behind most of his considerable accomplishments. He remained adventurous in both his business and private life (the racing cars were replaced by horses and he rode with the local Badsworth hunt) but the decisions he made on behalf of the company were based on good sense and sound judgement.

During his apprenticeship with the company, he had gained valuable knowledge, and had even written a book entitled *Gearing for Beginners,* all of

LEFT:
Young David Brown was appointed managing director of David Brown & Sons in 1932 when he was only twenty-eight years old. The executive offices at Park Works were sumptuously finished in oak panelling supplied by Waring & Gillow of London.

which stood him in good stead for running the business. He had become an experienced 'hands-on' engineer in his own right and never expected anyone to do anything that he could not do himself.

Rivers Fletcher, a former motoring colleague of David's, who had got to know him while Park Works was making gears for Raymond Mays's ERA racing car, provides the following description: 'Small in stature, reserved and modest, he was not easy to know. At first I found his apparent lack of enthusiasm in comparison with the effervescent Raymond Mays difficult to understand. But the hard Yorkshire business had given him a good engineering background and a very careful approach. ... A great asset was his choosing his managers so well.'

The young managing director's ability to choose the right man for the right job was soon evident. He recognised the importance of skilful and sustained marketing, and appointed Allan Avison as the firm's first sales manager. Avison had originally joined Park Works as a lad in 1911, becoming machine-shop foreman before moving into the administrative side of the business. His commercial ability proved to be a valuable asset to the company and he would later rise through the ranks to become eventually deputy managing director of the David Brown Corporation.

Mention must also be made of Arthur Sykes, who served the David Brown organisation for over fifty years and was responsible for many of the company's most important gear developments and technological achievements. Joining David Brown & Sons at the age of sixteen in March 1905, he successively held the posts of chief engineer, works manager, engineering controller and technical director, becoming recognised as one of the world's leading experts on gear production. He took special interest in the research department at Park Works and was affectionately known by the technical staff as 'Uncle Arthur'.

Working under Sykes through the 1930s was a pair of gifted research engineers, Dr Henry Merritt and Dr William Tuplin. Merritt's field of expertise was in transmissions, while Tuplin specialised in involute gear technology. Both men were brilliant, but a clash of personalities evidently resulted in great antipathy between the two. In 1937, Merritt left to work for the War Office and Tuplin was appointed chief engineer at Park Works, taking over full responsibility for the research department. During the 1950s, Tuplin accepted the post of Professor of Engineering at Sheffield University and later wrote a series of acclaimed books on British railway locomotives.

New Opportunities

Within just a few short years, young David Brown had revitalised the company on a

BELOW:
The Wallis & Steevens Advance was designed for silent running and was one of the first steamrollers to have precision gearing. The machine-cut gears and steel shafts in its transmission were manufactured by David Brown at Park Works.

ABOVE:
The Radicon assembly department at Park Works. Introduced in 1933, the Radicon worm-gear unit incorporated a cooling fan to dissipate the heat when running. It was an extremely successful product that had many applications and sold worldwide.

broader base, ready to move forward and seize new opportunities as they arose. The explosion of interest in motor transport brought a string of orders for spiral-bevel and worm-drive rear axles, differential units and steering gears, and a range of commercial vehicle gearboxes was added to Park Works' product line in 1936.

A good example of the many hundreds of early vehicles on Britain's roads that benefited from David Brown technology was the Advance steamroller, introduced in 1923 by Wallis & Steevens of Basingstoke in Hampshire. Steam traction engines are not usually recognised for their precision gearing and their movement is characterised by the rumble of their crude cogs. When planning the Advance, Wallis & Steevens were determined to design a steamroller to meet the requirements of modern methods of road construction. It had to be more efficient to compete against the latest motor rollers and was fitted with a double-cylinder balanced engine giving a very even torque and a quick-reverse feature.

The engine was matched to an entirely new and more modern transmission incorporating a two-speed gearbox and a four-pinion differential. The components for the transmission, including machine-cut steel spur-gears and machined-steel

RIGHT:
During the 1930s, David Brown & Sons built gearboxes, steering mechanisms and propulsion units for land, sea and air, including this six-speed transmission which was developed at Park Works for an early diesel-electric railway locomotive.

intermediate-shafts, were manufactured at Park Works by David Brown & Sons. It was a high-quality transmission designed for silent running.

One of the most significant of all David Brown's gear developments was the launch of the Radicon unit in 1933. The Radicon was a worm-gear reduction unit incorporating a cooling fan to disperse the heat created while running. It was totally enclosed within a ribbed casing, which served to further dissipate any heat build-up. A team working under Arthur Sykes was responsible for the design and the Radicon name was derived from the unit's ability dissipate heat through radiation, conduction and convection.

The Radicon had many applications and sold worldwide. Over 250,000 had been made by 1960 and production was eventually transferred to a new dedicated factory in Sunderland in 1962. The one millionth Radicon worm-gear unit was delivered in 1968 and the unit is still in production today. One of the most famous applications remains a unit that was installed to drive the revolving restaurant on London's Post Office Tower in 1965.

Returning to the 1930s, the decade saw David Brown & Sons manufacturing commercial vehicle

RIGHT:
David Brown gear products had many applications. This spiral bevel unit transmits the drive from a giant six-cylinder Ruston-Hornby diesel engine, turbocharged to 650hp, to a Gwynne drainage pump that can shift 5.8 cubic metres of water a second. It is one of three installed at Pode Hole pumping station, near Spalding in Lincolnshire. They are still in regular use and form part of the area's important flood defence systems.

and tank gearboxes, marine propulsion units, steering mechanisms for airships (including the ill-fated R-101) and liners, final-drives for railway locomotives and every trolley-bus built in Britain, as well as gear units for the ventilating fans for the new Mersey Tunnel. On top of this, the company's core gear and machine-tool activities were supplying African gold mines, Russian rolling mills, pumping stations, steelworks, paper and rubber mills all over the world. New overseas links were forged and, significantly, Mr David Brown was planning to diversify into a new 'end product' in the form of a farm tractor.

ABOVE:
A Huddersfield Corporation trolley-bus. David Brown claimed that every trolley-bus built in Britain incorporated its worm-drive rear axle. Huddersfield introduced its service in 1936 and there were 140 electric trolley-buses operating around the town by 1940.

CHAPTER 2 — *Ferguson and Brown*

A consignment of Ferguson-Brown tractors and implements outside the Park Works offices. The Scammell Mechanical Horse and Karrier Cob semi-trailer units were operated by the LMS railway company and were used to take the farm machinery to the freight depot in Huddersfield for dispatch by rail.

RIGHT:
Harry Ferguson, whose early experiments led to the development of the three-point linkage and a system of hydraulic depth control.

To understand fully David Brown's involvement with farm tractors, we must travel back in time to Ireland during the First World War where a young engineer by the name of Harry Ferguson was instructing farmers on behalf of the Irish Board of Agriculture during the 1917 ploughing campaign. The farmers were being taught the correct operation of their tractors and ploughs to improve their efficiency and standard of work.

Ferguson was not a farmer but he had been raised on a farm in County Down. At the time of the ploughing campaign, he was only thirty-three years old, but already had a great deal of engineering and commercial experience behind him and owned a respected motor business in Belfast. He was a flamboyant character with something of a reputation of a showman. His earlier exploits, racing cars and motorcycles, together with a successful attempt at making the first powered flight in Ireland in an aeroplane that he had designed and built himself, had brought him both fame and notoriety.

The position with the Irish Board of Agriculture had been offered after Ferguson, helped by his able assistant, Willie Sands, had staged a series of impressive ploughing demonstrations with an Overtime tractor and Cockshutt plough. The Overtime had recently been added to the list of agencies for leading British, American and French cars that Ferguson already held.

RIGHT:
A Ferguson-Brown tractor outside Harry Ferguson's birthplace, Lake House, near Dromore in Northern Ireland.

The early tractors were slow, heavy, cumbersome and sometimes unstable. Ferguson's experiences with them were to have a lasting effect on him. Being both a son of the soil and an engineer, the failings of these rudimentary machines were obvious to him; he questioned their efficiency and maintained that there must be a better way of connecting the plough to the tractor than dragging it along by a length of chain.

After the ploughing campaign was over, Ferguson returned to Belfast determined to mount the plough on the tractor in such a way that the two became one integral unit. His first plough, dubbed the 'Belfast' plough, was fitted to an Eros tractor conversion of a Model T Ford car and demonstrated across Ireland and England during 1918. This was followed by ten years of experimentation with plough designs, linkages and systems to control the depth of the implement in work.

Ferguson's ideas were translated into metal by Sands, a brilliant engineer in his own right, who was assisted by a pattern maker, Archie Greer. The culmination of their work was a lightweight mounted two-furrow plough that was attached to the tractor via a three-point linkage. The linkage was operated by a hydraulic lift that incorporated a draft control system to control the depth of the implement automatically.

After a prototype version of the lift and linkage was successfully tried out on a Fordson tractor in 1928, Ferguson began to approach a number of manufacturers with his system: Henry Ford gave him a lukewarm reception and appeared unenthusiastic. Allis-Chalmers in the USA, Ransomes & Rapier of Ipswich, Morris Motors and the Rover Company showed plenty of interest but little commitment. In the end, Ferguson realised that the only way forward was to build his own tractor.

The Ferguson Black Tractor

Ferguson's small engineering team of Sands and Greer, assisted by John Chambers who had served his engineering apprenticeship with Harland & Wolff, began drawing up plans for the tractor during 1932. The design of the tractor was loosely based on the Fordson, copying its simple unitary construction with the engine, gearbox and transmission forming the backbone of the machine.

Parts were either handmade in Ferguson's workshops or subcontracted to outside suppliers, with several of the castings made from aluminium

LEFT:
Plans for Harry Ferguson's prototype 'Black Tractor' were drawn up during 1932 and the completed machine was ready for trials the following year.

RIGHT:
The prototype 'Black Tractor' was powered by an American Hercules engine. David Brown & Sons were among several subcontractors involved in the design with Park Works supplying many of the transmission, rear axle and steering box components.

alloy to keep the weight of the tractor as light as possible. The engine came from America and was supplied by the Hercules Corporation.

Most significantly, in 1933, Harry Ferguson placed the order for the three-forward and single-reverse speed gearbox, the spiral-bevel crown wheel and pinion for the rear axle, and the steering box with David Brown & Sons at Park Works. The Huddersfield firm was evidently chosen for the work because Ferguson had spent time on a farm at nearby Honley back in 1918 while demonstrating his 'Belfast' plough in Yorkshire.

The newly completed prototype tractor was painted black and was ready for trials by the middle of 1933. It incorporated Ferguson's lift, linkage and hydraulic draft control system, powered by a four-cylinder plunger-type pump with an aluminium body and cast-iron pistons and cylinders. The pump was driven off an extension

RIGHT:
The original 'Black Tractor' photographed after being restored in Ferguson's workshops in Coventry in 1950. Today, the machine is preserved in the Science Museum in London.

of the gearbox layshaft, which meant that the hydraulics only worked when the tractor was moving.

The pump sucked oil from the barrel housing and passed it through a simple control valve to an internal ram cylinder that raised the lift linkage. It was a single-acting system and the implement had to rely on gravity and its own weight to lower itself back into work. A single lever on a quadrant to the right of the driver's seat operated the lift via an internal linkage to the control valve.

The hydraulic system also controlled the working depth of the implement in relation to the contours of the land by 'sensing' the changing draft present between the soil and the implement. The depth was governed by a sensing mechanism connected to the control valve, receiving signals through the lower-link arms. Field trials showed up some inadequacies of the hydraulics, but further experimentation led to the adoption of top-link sensing that greatly improved the system. Other refinements included moving the control valve from the pressure to the suction side of the pump to counteract problems with the oil overheating.

Once Ferguson was satisfied with the design of his tractor, he organised a series of demonstrations in Ulster and England. One of these was arranged in Huddersfield on land close to Park Works. Being fascinated by motor vehicles of any kind, Mr David Brown attended the demonstration with great interest and was evidently highly impressed by both Ferguson and his 'Black Tractor'. Having recently toured several tractor plants during a trip to the USA, Brown was well aware of the impact that mechanisation was having on agriculture, but was equally conscious that many of the large and heavy American tractors were unsuitable for British farms.

Ferguson's tractors were different; his principles were simple but the advantages were manifold. The hydraulic system and its ability to lift, carry and control the implement made the tractor both more adaptable and more manoeuvrable than any other machine on the market. Because the three-point linkage transferred the weight of the implement onto the rear wheels, the tractor needed no added ballast and was much lighter than its contemporaries. The converging links converted the natural forces affecting the implement into a strong forward and downward thrust and made the tractor very stable, while ball joints at the end of the lower-link arms and top-link allowed the implements to be quickly and easily attached.

Mr David Brown was drawn in by Ferguson's infectious enthusiasm, partly because of their mutual interest in motor racing, and had convinced himself that the tractor project was worth consideration when he heard that the Craven Wagon and Carriage Works, a subsidiary of the Sheffield steel company, Thomas Firth and John Brown, had already agreed to take on its manufacture. He had no need to worry for this was a short-lived arrangement. Harry Ferguson was undoubtedly a gifted and clever individual but he was a very difficult person to work with and liked everything his own way. It was not long before the Sheffield company began to regret its decision and was more than relieved to hand over the manufacturing agreement to David Brown after the negotiations with Ferguson broke down in 1935.

The initial agreement was for Brown to build Ferguson 750 tractors. The agreement was evidently biased in Harry Ferguson's favour as the Ulsterman reputedly claimed, 'When I sell them, I will pay for them'. Several David Brown personnel referred to this as 'The Ferguson Trick'; it was a trick he was later to repeat with Henry Ford.

The Ferguson Type A

Securing the agreement to manufacture the Ferguson tractor may have pleased young David Brown, but his father, Frank, was not convinced that it was a good move or a good deal for the company. Furthermore, he could see no future in manufacturing motor vehicles, having made little money out of the earlier Dodson and Valveless car ventures. He also felt that the timing of the tractor project was inopportune because the company had just been involved in a lot of expense in developing the Penistone foundry. Furthermore, Frank and David had just committed much of their own personal money to consolidating the family's position on the board. They had acquired the late Percy Brown's shares and had recently bought up the majority of David Brown & Sons shares that had been originally issued to the former Jackson directors at the time of the acquisition of P. R. Jackson Ltd in 1930.

However, David remained resolute and

RIGHT:
Harry Ferguson at the wheel of his Type A tractor. The manufacturing agreement for the machine had been secured by David Brown in 1935.

BELOW:
The Ferguson Type A was manufactured by David Brown Tractors Ltd. and sold and serviced by Harry Ferguson Ltd. It was based on the design of the prototype 'Black Tractor' with a few modifications.

determined; he was not really interested in tractors for their own sake, but was looking for an end product for the company. He was genuinely concerned that, as gear manufacturers, David Brown & Sons' future growth would be restricted if it remained forever a subcontractor, always relying on supplying parts and components for other people's machines. He strongly felt that an end product would give the company greater stability and a tractor was ideal because it was a machine that contained a lot of gears.

While not offering his son much encouragement publicly or at board level, Frank Brown privately supported David's ambitions and helped him borrow some capital and set up David Brown Tractors Ltd as a separate subsidiary. At the same time, he agreed to the parent company taking a small financial interest in the new concern and rented David space

LEFT:
Harry Ferguson puts the Type A tractor through its paces. His partner, Mr. David Brown, was keen to be involved in the tractor project in order to give his gear manufacturing business a viable end product.

to build the tractors in a former pattern shop that had recently been vacated at Park Works.

David Brown Tractors Ltd was only responsible for manufacturing the tractor. Ferguson handled the sales, marketing, distribution and servicing. He traded as Harry Ferguson Ltd, occupying the former Karrier Motors offices in Cable Street at Longroyd Bridge, just north of Lockwood. Karrier, a local commercial vehicle manufacturer dating back to 1904, had recently closed down its Huddersfield factory in nearby St Thomas Road and moved its operation to Luton.

A few modifications were made to the 'Black Tractor' design before it was put into production; revisions were made to the hydraulic system and

BELOW:
Six pre-production tractors for demonstration, publicity and training purposes were hand-built at Park Works during the spring of 1936.

RIGHT:
The launch of the Ferguson Type A tractor at Park Works on 19 August 1936 was followed by a two-hour field demonstration held at a nearby farm.

linkage, and the clutch pedal was moved from the right to the left side of the barrel housing. David Brown's engineers had their own suggestions, but these were almost always overruled by Ferguson who remained characteristically autocratic; the tractor was his baby and only he would change it. Arthur Sykes was in overall charge of Park Works' responsibility for engineering matters relating to the tractor, and represented David Brown on the design team.

During the spring of 1936, around six pre-production tractors for demonstration, publicity and training purposes were hand-built by David Brown fitters in one of the machine shops at Park Works because the building earmarked for tractor production had not yet been cleared. The assembly team was supervised by the workshop foreman from the automobile gear section, Harry Pilkington, and led by one of the fitters, Fred Armitage, with Harry Ferguson watching their every move. Considering that David Brown was the senior organisation with a lifetime of engineering excellence already behind it, it is unlikely that Park Works' personnel were comfortable with this constant surveillance. There was undoubtedly friction between the two camps, and the engineering staff at Park Works allegedly referred to their counterparts at Cable Street as the 'Ferguson Mafia'.

The new tractor, known as the Ferguson Type A, was presented on 19 August to influential figures from the world of agriculture and representatives of the national and trade press. The launch began with a tour of the tractor manufacturing plant at Park Works, followed by an address given by Harry Ferguson in which he explained the virtues of the new tractor and how its hydraulic system worked. After lunch, the guests were ferried by coach to a nearby farm for a two-hour field demonstration.

The launch was timed to coincide with the release of the first press announcements heralding a 'new British tractor'. It was an important milestone in the history of farm mechanisation and marked the introduction of the first production tractor to incorporate a hydraulic lift and a converging three-point linkage.

The tractor was painted battleship grey and was built using heat-treated high-tensile alloy steel, with the housings for the clutch, hydraulic pump and transmission cast from aluminium alloy. It weighed only 16½ cwt to fit in with Ferguson's belief that a tractor should be no heavier than a single farm horse.

Unlike the prototype 'Black Tractor', the Type A was powered by a British-built engine supplied by Coventry Climax Engines Ltd of Widdrington Road, Coventry. The company was a long established manufacturer that specialised in 'off-the-shelf' units for a variety of industrial and automotive applications. It had a number of licensing agreements with American firms in place and was currently supplying a British version of the Lycoming petrol engine to the Gilford Motor Company, a lorry and bus maker based at High Wycombe in Buckinghamshire.

The 20 hp Coventry Climax E Type engine specified for the production Ferguson tractor was based on the Hercules IX B unit that had powered the 'Black Tractor'. No doubt some sort of licensing agreement between Coventry Climax and Hercules existed, although at least one former Ferguson employee believed that the E Type was not officially sanctioned. However, it does seem unlikely that Coventry Climax would risk its reputation by making an unauthorised copy of the American engine.

The British and American engines were nearly identical to one another, with the same 3¼ in. bore with a 4 in. stroke and a capacity of 133 cu in. (2,175 cc). Some Coventry Climax engines were

ABOVE:

The 20 hp Coventry Climax engine as fitted to the first production Ferguson Type A tractors. This early power unit, known as the E Type, can be identified by its 'paired-plug' head and shallow sump. The carburettor was supplied by Solex.

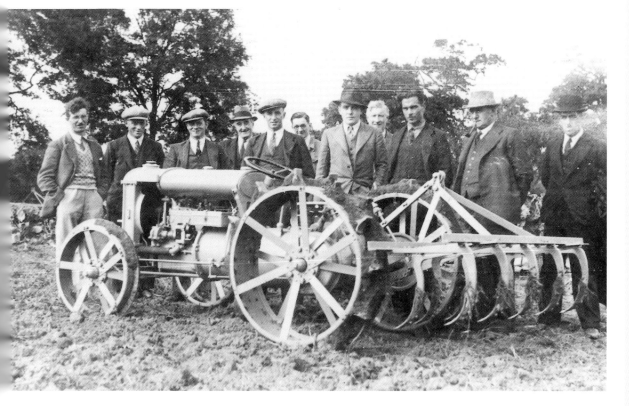

LEFT:

The Ferguson Type A cost £224 with full hydraulic lift and linkage. The tractor's main castings were made from aluminium alloy and it weighed just 16½ cwt.

RIGHT:
The new Ferguson tractor was available with a range of four implements. The Type B plough was a two-furrow model that had been developed from Harry Ferguson's earlier experiments.

even fitted with Hercules valves, and the only real difference was that the British power unit had a Solex carburettor. Incidentally, later versions of the Type E were fitted to wartime generating sets and fire pumps while legend has it that the same power unit provided the starting point for the development of Coventry Climax's famous racing car engines of the late 1950s and early 1960s.

The Type A tractor was priced at £178 without hydraulics. If you wanted to make full use of the Ferguson System, a model with full hydraulic lift and linkage cost £224. Four implements costing £26 each were launched with the tractor in 1936. They were all assembled at Park Works with frames made from heat-treated alloy steels.

The plough, designated the Type B, was a two-furrow model, loosely based on earlier Ferguson designs. The Type C was advertised as a 'general

RIGHT:
The Ferguson Type C general cultivator had seven spring-loaded tines and was designed to override any obstructions such as stones or tree roots.

cultivator' and had seven spring-loaded tines to override any obstructions, while a rowcrop cultivator with nine rigid tines was listed as the Type E. The Type D designation was reserved for a three-row ridger that had a steerage fin to keep it in a correct line behind the tractor when working on slopes or splitting ridges. A single-furrow digger plough with a 16 in. mouldboard was added to the range at a later date.

Early press releases for the Ferguson tractor claimed that the Coventry Climax engine would burn paraffin or vaporising oil when warm. However, no vaporiser attachment was available for the early models and most tractors ran on straight petrol. The Type A did have a twin-compartment fuel tank, but the smaller compartment was only a two-gallon reserve for the main eight-gallon petrol tank.

By September 1936, the Ferguson had been demonstrated as far afield as Ballyclare in County Antrim, Dormington Court Farm in Hereford and the Channel Islands, and Harry Ferguson Ltd claimed to have dispatched fifty machines. The first to be sold, tractor No.12, was bought by John Chambers's father in June 1936.

ABOVE:
The Type E cultivator had nine rigid tines which could be arranged for rowcrop work. All the Ferguson implements cost £26 each and their frames were made from heat-treated alloy steels.

LEFT:
The three-row ridger was known as the Type D. It had a steerage fin that helped eliminate any lateral movement of the implement and kept the ridger in a correct line behind the tractor when working on slopes.

Walter Hill, an enigmatic character who was well versed in the tractor business and had variously held positions at Muir-Hill, Rushton, Roadless and Bristol Tractors, was engaged as general manager of Harry Ferguson Ltd. The sales department, which operated out of the offices in Cable Street, was managed by Harold Willey, assisted by Trevor Knox, a former Ferguson apprentice from Belfast, who handled sales in Scotland and Northern Ireland. A former army officer, Captain Arthur Arlborough-Smith, was also appointed to act as company secretary.

John Chambers ran the service and training department, which was established nearby in the old Karrier factory. Fred Meadowcroft, an experienced young mechanic, was recruited from A. Hawke & Sons, the local Commer dealership at Folly Hall, to act as field engineer for the Ferguson tractor and was provided with an Austin 10 service van. The demonstration team consisted of Bob Annatt, Cyril Marshall and Bill Harrison.

Once production proper was underway, Harry Ferguson would visit the assembly line at Park Works at least once or twice a week. Len Craven, who was appointed storekeeper in charge of the tractor and implement parts, remembers him as a small, white-haired man with a slight stoop, always accompanied on his weekly inspections

RIGHT:
John Chambers, who ran Ferguson's service and training department, demonstrates the Type A tractor to a group of young farmers. The new machine was regarded a something of a novelty but initial sales were disappointing.

by his 'three henchmen': Sands, Greer and Chambers. Mr David Brown had evidently already got the measure of the man, and judiciously stayed away during Ferguson's visits, leaving Arthur Sykes to escort the Ulsterman down the line.

Tractor production at Park Works began very slowly. Like many other large engineering concerns, the plant operated on a system of sanctions to control manufacturing, whether it be for tractors, gear units or any other component or sub-assembly. Before anything was put into production, a building order or sanction had to be raised, dictating in advance the number of units to be built to regulate the procurement of raw materials and parts. Much to Harry Ferguson's dismay, the David Brown hierarchy decreed that the first tractor sanction was for 100 machines only. The Huddersfield company was being prudent and wanted to judge farmers' reactions to the new tractors before investing too much in the venture, while Ferguson typically believed that the best way to increase sales was by saturating the market.

Unfortunately, the Ferguson tractor's reception was disappointing and overall sales were slow. It was feted by the press and engineering circles, possibly because it was something of a novelty. However, novelties are often viewed with suspicion and many farmers resisted buying a machine that needed its own special implements and was nearly $80 dearer than the tried and trusted Fordson. Sales were also limited by the lack of a power take-off, vaporiser, pneumatic tyres and a choice of wheel equipment - matters which Harry Ferguson refused to address, arguing that the tractor was designed for the small farmer who had no need for such features.

Ferguson-Brown

The first sanction of 100 tractors had been built by the end of 1936, and the second sanction, this time for 250 machines, had been completed by the following summer. It was at about this time that the fledgling tractor venture faced its first hurdle when Coventry Climax announced that it could no longer continue to supply small batches of engines just for the tractor. The company had other more important commitments, in particular a large contract from the Ministry of Supply for Godiva fire pumps. The problem was resolved after Coventry Climax agreed to sell David Brown its casting patterns for the Type E power unit. Engine manufacturing was transferred to Park Works, while the castings were made at Penistone.

Nobody seems to know exactly at what stage

LEFT:
An early Ferguson tractor with the Coventry Climax engine. Nearly 350 machines had been built by mid-1937. Note that the tractor is fitted with the improved Burgess air-cleaner.

BELOW:
By the time 500 Ferguson tractors had been built, a changeover had been made to engines manufactured by David Brown. Features of the new power unit included a deeper stepped sump, a slightly smaller bore and a different head with equally spaced plugs. This tractor also has the early oil strainer with the brass handle that turned the filter against a brush.

the David Brown engines were introduced. Part changes brought in at tractor No.256 during the second sanction seem to suggest that this marked the beginning of a phased changeover. There appears to have been a period of transition and it is generally accepted that around 500 tractors were built before Coventry Climax engines were completely phased out. During this transitional period, both makes of engines were used, and it seems that some David Brown blocks were fitted with Coventry Climax heads, and vice-versa.

The David Brown engine was only slightly altered from the Coventry Climax unit, the main changes being that it had a 3⅛in. bore as well as a slightly different cylinder head and a larger sump. The Donaldson dry air-filter fitted to the early tractors was replaced by a more efficient Burgess oil-bath air-cleaner. Revisions were also made to the carburettor.

A problem with oil starvation, which had led to a number of main bearing failures on some of the early engines, was also addressed. The root of the problem had been a pipe that fed oil to the timing case. It tended to pick up sludge from the timing gears, which blocked the oil lines and led to starved bearings. A pencil filter was fitted to the feed pipe, but this also clogged up and made matters worse. Fred Meadowcroft, who had first identified the problem in the field, suggested a modification that did away altogether with the feed to the timing case.

A crude but effective oil-strainer with a self-cleaning mechanism introduced on the David Brown engines was a further improvement as was

RIGHT:
A consignment of Ferguson tractors and implements leaving Huddersfield by rail. David Brown was very careful to regulate production while demand for the tractor remained static.

BELOW:
The Ferguson Type A tractor was dogged by reliability problems, in particular component failures caused by stress fractures to the aluminium alloy castings.

the larger sump. The shallower sump fitted to the earlier engines had allowed the oil pump to become starved of oil when working on slopes. The fact that around 10 percent of the early tractors had already had their engines replaced under warranty due to bearing failure underlined the need for these timely modifications.

During 1937, Mr David Brown began to assemble his own tractor design team led by Albert Kersey - a gifted engineer who would later prove pivotal in the company's tractor developments. Albert Henry Kersey, to give him his full name, was born in 1900 at West Ham in North London. Before joining David Brown, he had previously been employed by the Gilford Motor Company.

Gilford had been wound up in 1937 after struggling with outdated designs powered by the thirsty American Lycoming petrol engines. Kersey was recruited by David Brown & Sons as a design draughtsman and was given the responsibility of product manager for the worm-gear section at Park Works before Mr David Brown decided that his automotive experience would be put to better use in the tractor division.

Kersey's team went on to suggest a number of further improvements to the Ferguson

ABOVE:
The Ferguson service engineers were kept busy sorting out warranty claims and at least 10 percent of the engines had to be replaced due to bearing failure. This Austin 10 service van was operated by field engineer, Fred Meadowcroft. Note the new Ferguson-Brown company name on the van door.

tractor, including fitting a more powerful engine because they felt its miserly 20 hp was not enough. Fred Meadowcroft recalls being shown three Meadows engines that had developed bearing problems and believes that the original intention had been to use these engines in the Type A tractor.

However, Harry Ferguson ignored Kersey's advice. He continued to adamantly oppose every little change, even though sales remained static and his distribution company was running into financial difficulties. For a time, demand for the tractor slightly exceeded supply, but only because David Brown was very careful to regulate production. Ferguson's exhortations for increased production fell on deaf ears and the third sanction was once again restricted to 250 machines.

David Brown was also not making any money out of the tractors and furthermore was saddled with a number of expensive warranty claims due to component failures caused by stress fractures to the aluminium alloy castings. A number of broken transmission housings, which saw the tractor literally split in two, caused the company particular embarrassment. Some of the castings did not even get as far as the assembly line and fractured while being unloaded from the delivery lorry in the works.

The transmission housings, dubbed 'milk churns' by the assembly line fitters because of their distinctive tapered aluminium appearance, formed the main backbone of the tractor. As an interim measure, Albert Kersey developed a cast-iron housing while different specifications of high-grade alloys were investigated. An expensive cast-aluminium alloy known as RR50, usually used for racing motorcycle cylinder heads, was eventually

ABOVE:
Ferguson-Brown Ltd was formed on 25 September 1937 and personnel from the company pose with one of the tractors. Those who have been identified include (1) David Brown (joint managing director); (2) Fred Meadowcroft (field engineer); (3) Albert Kersey (tractor design); (4) Godwin Atkinson (tractor assembly fitter); (5) Harold Thompson (works manager); (6) Jack Layton (tractor assembly foreman); (7) Donald Kilburn (manager of tractor and parts dispatch); (8) Miss May (Walter Hill's secretary); (9) Walter Hill (general manager); (10) Bob Annatt (chief demonstrator); (11) Fred Armitage (tractor parts assembly foreman); (12) Bill Harrison (demonstrator).

specified for the housings, but not before some hydraulic top-covers had also been fabricated in cast iron. The alloy castings were supplied by the Birmingham Aluminium Casting Company of Smethwick in Staffordshire in batches of twelve at a time.

Major problems with differential and hydraulic pump failures also had to be addressed and at one time there were four service engineers in the field attending to complaints. Unfortunately, a series of warnings for the operator contained within the Ferguson instruction book only served to further highlight the tractor's inherent weaknesses. Eventually, it became the custom of many dealers to take the book out of the toolbox and destroy it before the tractor was handed over to the customer following an incident where one farmer refused to take delivery after reading all the 'don'ts' and restrictions in the manual.

The lack of confidence in the Ferguson tractor affected the dealers, and three of the first distributors appointed were virtually driven out of business. One of these, Cyril Ratcliffe of Colchester, was so incensed by the whole affair that he wrote a letter of complaint to Walter Hill on toilet paper, explaining that he did not want to waste good stationery on the matter!

With both sides of the tractor venture losing money, the recriminations and dissension between the two parties came closer to the surface, but Brown had greater financial backing than Ferguson and was able to manoeuvre the Ulsterman into merging his sales company with David Brown Tractors Ltd. The new concern, known as Ferguson-Brown Ltd, was formed on 25 September 1937. Harry Ferguson and David Brown were joint managing directors, but Ferguson held only a minority shareholding; he had to accept Arthur Sykes as technical director and found his influence on sales and service greatly diminished. As a result of the merger, several of the Ferguson staff, including Walter Hill, Captain Arlborough Smith, Fred Meadowcroft, Bob Annatt, Cyril Marshall and Bill Harrison, became David Brown employees. The new company title became a familiar nickname for the tractor, which even today is still better known as the 'Ferguson Brown'.

The Tractor Assembly Shop

Tractor production at Park Works was located in a five-storey edifice known simply as the 'concrete building', which was easily identified by the large 'Ferguson-Brown' sign on its roof. Built in 1916 as a three-storey structure, it was one of the first pre-stressed concrete buildings in Britain. The top two floors were added in 1928. The building was located opposite the heavy-machine section of

LEFT:
An aerial photograph of Park Works in the immediate post-Second World War period. Note how the five-story concrete building that was used for tractor production from 1936 to 1939 dominates the site.

the right of the service road that ran from the stone archway over the main gate. Now turned into an office block, it still dominates the view from the No.1 Gate at Park Works.

Seventy years ago, the ground floor of the concrete building, at times used as a canteen, became home to a new commercial vehicle gearbox line for the three-wheel Scammell Mechanical Horse. Four-speed gearboxes for these light three-wheel articulated units and their Scarab and Townsman successors were made at Park Works into the mid-1960s. The remaining four floors of the building were used as a pattern shop and store until the foundry work and most of the patterns were moved to Penistone in 1936. The first, third and fourth floors were then cleared and assigned to tractor production.

The main tractor assembly area was on the first floor. Here, the main components, including the hydraulic unit, gearbox, rear axle, brake and steering sub-assemblies, were built up at one of five separate stations, each with its own workbench. In front of these was the final assembly line that had the capacity for five tractors. Once the engines were

LEFT:
Photographed in October 1995, the original Ferguson-Brown assembly building at Park Works still dominates the view from the No.1 Gate. The concrete structure was built in 1916 with the top two floors added in 1928.

RIGHT:
The main tractor assembly area on the first floor of the concrete building at Park Works. Sub-assemblies were built up at five separate stations, each with its own workbench, behind the final assembly line.

buckled up to the transmission, the units were run in for an hour by an electric motor drive. Then, after an oil change, the tolerances were checked and bolts tightened before the engines were tested for full power on a Heenan & Froude hydraulic brake dynamometer.

The third floor was the implement assembly shop where the ploughs, cultivators and ridgers were put together from components mainly supplied by outside subcontractors. The top floor housed the paint section, which was quite automated for the time with a system of conveyors

RIGHT:
A plan of the main tractor assembly area at Park Works.

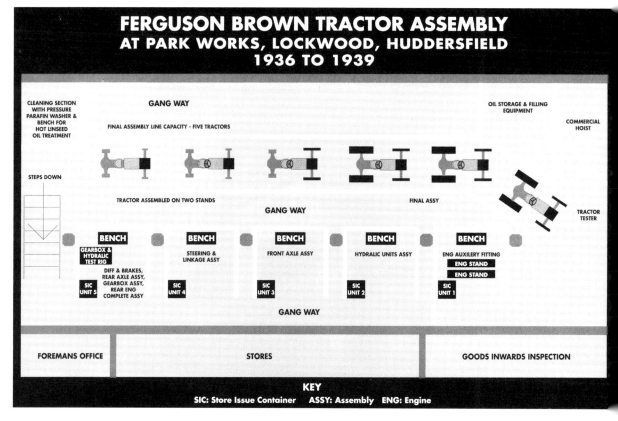

LEFT:
The implement assembly shop on the third floor of the concrete building at Park Works. The manufacture of the components for the ploughs, cultivators and ridgers was subcontracted to outside suppliers.

BELOW:
The machine shop for the tractor works was housed in a large ground-floor annexe to the concrete building and was equipped with the latest machine tools. The total floor space allocated to tractor production at Park Works totalled an acre in area.

and turntables. The painting booths had fume extraction fans and the paint sprayers were issued with gas masks – protection not normally considered necessary in the 1930s. All the tractors and implements received two coats of paint with a special 'rust-proofing' brown primer followed by a grey topcoat. This floor was also a storage area both for completed tractors, awaiting final inspection prior to being dispatched to Cable Street, and for components such as radiators for the assembly line.

Behind the concrete building and at a slightly lower level was a large machine-shop annexe. It had previously been used for manufacturing both worm gears and the bronze steering mechanisms for Atlantic liners until being reorganised as the machine shop for the tractor operation. With such a high reputation for engineering excellence to uphold, David Brown spent enormous sums of money in equipping the shop with the very latest in modern machine tools, including multi-spindle drilling machines, jig-boring machines and no less than seven Herbert No.7A multi-automatic machines.

One machine alone, a Bullard 'Mult-au-matic' tool for machining gear blanks, cost over £7,000, which was a considerable sum in the 1930s. Another £6,000 was spent on setting up a bolt-making plant. All the bolts used on the tractors and implements were made from high-tensile

RIGHT:
The tractor engine line conveyor and machine shop at Park Works. Engine production was established on the second floor of the concrete building in 1937.

steel and were fitted with special cyanide hardened nuts. A cruciform four-head drilling machine prepared the rear-axle casing in one seventeen-minute operation. Probably slow and old-fashioned by today's standards, it was regarded as highly automated and was the last word in precision engineering at the time of the Ferguson-Brown.

The second floor of the concrete building, which for a time had remained a pattern store, was cleared for engine production after the David Brown power units were introduced in 1937. A new engine line conveyor was installed next to a new engine machine shop. Here, the raw castings were placed in a Lumsden grinder before passing through various drilling and boring machines. The block was prepared by a Kitchen & Wade cylinder-boring machine followed by an Archdale honing machine to give the bores their hardened finish.

At its height, the tractor assembly shop was turning out ten to fifteen machines per week. The total floor space allocated to tractor production in the building added up to an acre in area. Machines and materials were moved between the five levels by a platform hoist that was powered by line-shafting and had a reputation for being slow and unreliable.

The works manager was Harold Thompson. His office was on the first floor and he was assisted by Harry Pilkington who acted as works superintendent. The machine shop was run by Arthur Blackwell, who had joined Park Works as foreman of the bevel-gear cutting section in 1928 after nearly twenty years' experience with several leading motor firms. He was appointed machine shop superintendent in 1937.

Around thirty men were employed in the tractor assembly shop when it opened in 1936 but this number grew as production was stepped up and the engine line was added. Many of the fitters were former Karrier Motors personnel who had been made redundant after the lorry firm was absorbed into the Rootes Group in 1934. Karrier production had been moved to the Commer plant in Luton, which had led to the closure of its Huddersfield factory.

After the formation of Ferguson-Brown Ltd, the service and training department was also relocated from St. Thomas Road to the ground floor of the concrete building at Park Works.

Captain Arlborough-Smith brought in one of his former army colleagues, Bill Ketchell, to run the service school. Fred Meadowcroft took over the responsibility for the training school, teaching farmers, operators and dealers the principles of operating Ferguson equipment, while still acting as service engineer at the same time. Bob Annatt was appointed chief demonstrator and his team was augmented by one of the implement fitters, Ernest Kenyon.

The tractor operation also had its own parts store, which was next to the machine shop annexe behind the concrete building. The storekeeper, Leonard Craven, had joined David Brown & Sons as a temporary labourer in 1934 after working in a local worsted mill. As an example of the tradition of long service established by personnel in the tractor division, Kenyon, who eventually moved into sales, finally retired as product applications manager in 1975, while Craven retired as parts manager a year later. After forty years of dealing with tractor parts, the one part number that remains etched in Len Craven's mind forever is G3 - the infamous Ferguson transmission housing!

The Last Sanction

Once the Ferguson-Brown's early teething troubles had been ironed out, production at Park Works settled into a steady flow. A few detail changes and new accessories, planned in the main by Kersey's team and grudgingly accepted by Harry Ferguson, helped stimulate demand. Sales actually rose during 1938, buoyed up to a certain extent by the reputation of the Brown name incorporated into the sales company.

At Harold Willey's suggestion, the company concentrated on selective selling in areas of the country where the Ferguson-Brown was practically unknown and sales would not be dogged by the tractor's earlier poor reputation. A big sales drive was made in Scotland where fourteen tractors were eventually sold to one farm. The Ferguson was particularly popular in hilly areas because it offered greater stability when working on steep slopes. Export orders remained elusive, but an impressive demonstration of the Type A that John Chambers gave in Norway in the April of 1938 generated an immediate order for twenty-nine tractors to Scandinavia. The fourth production sanction, which had been set at 500 tractors, was increased to 750.

The Ferguson Type A was now offered with a choice of wheel equipment, including pneumatic tyres and narrow steel wheels for rowcrop work. Mudguards or fenders, described in the Ferguson-

LEFT:
The Ferguson-Brown tractor was popular in the hilly areas of the country where it offered greater stability for working on slopes. Targeted sales campaigns in Scotland led to increased sales.

RIGHT:
Pneumatic tyres and mudguards were offered for the Ferguson-Brown tractor from 1938. This unrestored example does not have the correct rear wheel centres but otherwise is in very original condition.

ABOVE:
A Ferguson-Brown tractor adapted for inter-row cultivation. Narrow steel wheels for rowcrop work were also available.

Brown catalogue as 'wheel guards', became available for the first time. A special hop and orchard model, with the front axle moved back 12 in. to allow a tighter turning circle, was introduced in May 1938, while a few industrial models with larger rear wheels and pressed-steel front wheels were sold to municipal authorities.

Other improvements instigated by David Brown's engineers included the introduction of a combined power take-off and belt pulley attachment, a conventional oil filter incorporating an oil pressure gauge, and a vaporiser to enable the tractor to run on paraffin.

The use of a vaporiser attachment was always a contentious issue as far as Harry Ferguson was concerned. He preferred the clean-burning simplicity of a petrol engine and disliked the use of the heavy distillates because they tended to dilute the sump oil. However, with the Second World War looming, petrol was in short supply and was becoming expensive. Farmers demanded a tractor that would run on the cheaper vaporising oil and even Ferguson had to agree to their requests.

The vaporiser for the Type A was devised by Arthur Gladwell of Gladwell & Kell Ltd, Grays Inn Road, London. Known as the Gladwell Kerosene Carburettor and Superheater, it was offered as a kit that included a temperature gauge and cost £15. The vaporiser became approved Ferguson-Brown equipment from November 1938, but was neither very efficient nor very successful.

Mr David Brown remained unhappy with the Ferguson-Brown and felt that its limitations far outweighed its obvious advantages. Not only did he think that the tractor needed to be more powerful, he also believed that it should be heavier and have the integral strength to handle trailed implements to appeal to those farmers who did not necessarily want to invest in expensive mounted equipment. Brown continually battled with Ferguson over the design, but the Ulsterman refused to compromise.

By the autumn of 1938, the rift between the two parties had deepened and there was almost no contact between the two camps. Harry Ferguson's visits to Park Works became increasingly infrequent as he returned to Belfast for long

LEFT:
A Ferguson-Brown with the Gladwell & Kell vaporiser attachment, which was supplied as a kit complete with temperature gauge. The tractor has non-standard David Brown wheel centres.

periods. During his absence, Mr David Brown had secretly instructed Albert Kersey to push ahead with plans to design a more powerful machine.

Unbeknown to Brown, Harry Ferguson had secret plans of his own and had gone to the USA with two Type A tractors to seek another audience with Henry Ford - this time with greater success. Following a demonstration held at Ford's home in Dearborn that October, an agreement was secured for the Ferguson System to be incorporated into Ford tractors. An elated Harry Ferguson returned to England to be confronted by a disgruntled David

LEFT:
Farmers who resisted buying the expensive Ferguson-Brown mounted equipment found that the tractor was not really heavy enough to cope with their trailed implements.

The David Brown Tractor Story

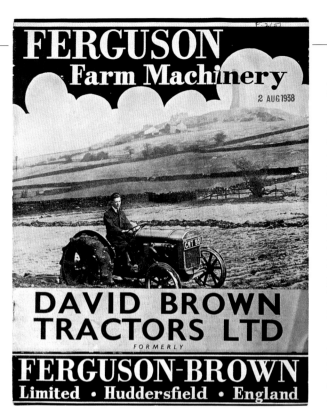

RIGHT:
Ferguson-Brown Ltd. was renamed David Brown Tractors Ltd with effect from 1 February 1939 and the sales brochures were overprinted with the new company title.

BELOW:
A Ferguson-Brown tractor in the company of another David Brown product - a 1950 Lagonda 2½ litre saloon. Just over 1,350 Ferguson Type A tractors were built before production ended in mid-1939.

Brown who had got wind of the American trip.

Initially, Ferguson denied that he had talked to Ford, but eventually details of the deal emerged. Although surprised by his erstwhile partner's actions, Brown was undoubtedly relieved to be rid of Ferguson and offered to buy out the Ulsterman's share in Ferguson-Brown Ltd. Ironically, the Ford and Ferguson partnership would also prove to be an unhappy relationship. It eventually broke up over recriminations that Ferguson's distribution division was taking a greater cut out of the resulting tractor than the Ford Motor Company who provided all the manufacturing facilities.

Mr David Brown did not often comment on his relationship with Ferguson, but was later to remark: 'Harry Ferguson was a man who could sell ice creams to Eskimos. He had tremendous charm when he cared to turn it on. In fact, I

ABOVE:
Some of the very last Ferguson-Brown tractors exhibited on David Brown's stand at the Royal Highland Show in June 1939.

consider him to be one of the world's greatest salesmen. On the other side of the penny, he could be one of the most difficult and awkward people it's ever been my task to come across.'

Ferguson-Brown Ltd became David Brown Tractors Ltd with effect from 1 February 1939. There were still nearly 100 tractors of the final sanction to be built. However, under the terms of the termination agreement with Ferguson, the renamed company had a non-executive and non-assignable licence for manufacturing, selling and servicing the last of these Type A models until production ended in mid-1939. No production figures are available, but it is assumed that 1,356 Ferguson-Browns were made. This is accounted for by the 1,350 total of the four sanctions plus the six pre-production models. The latest Type A tractor known to be in existence is No.1354.

David Brown had ambitious plans for the future with a new model in the pipeline and was negotiating the purchase of a site for a new dedicated tractor plant. The very last few Ferguson-Browns were to be eventually assembled at this new factory at Meltham. Advertisements placed in the agricultural machinery journals for July 1939 announced 'drastic reductions' in the price of the Type A in a final last-ditch attempt to move the remaining stocks of the old model. The tractor on steel wheels complete with the hydraulic unit was reduced to £198. Pneumatic tyres cost an extra £47.

The same magazines also carried artistic impressions of 'the new David Brown tractor', which was launched that month and was first seen at the Royal Show in the company of three Ferguson-Brown tractors. It was not so much a case of out with the old and in with the new as out with the grey and in with the hunting pink.

CHAPTER 3 Swords before Ploughshares

A David Brown tractor is used to clear rubble from the basement of the Aero Block at Meltham Mills following a devastating flood in 1942. The basement housed the heat-treatment plant and there was some urgency to get it operating again so as not to interrupt vital war production. Note the security guards on the main gate, which led into the plant off Meltham Mills Road.

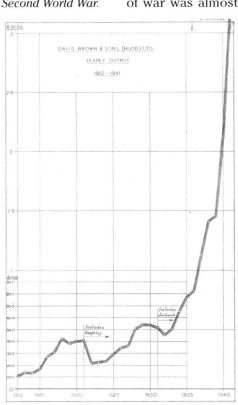

RIGHT:
A tractor demonstrator dons his gas mask as David Brown prepares for war. Although the company pressed ahead with plans to introduce the new tractor, its eventual production had to be curtailed in favour of defence contracts.

BELOW:
A graph of the yearly output of David Brown & Sons from 1912 to 1941 showing how defence contracts and munitions production dramatically increased the company's turnover in the lead up to and early years of the Second World War.

We will never really know what kind of impact the new David Brown Tractor would have had on the market because its launch and eventual production were both affected and governed by the gathering clouds of war. Even while the tractor was still in its period of gestation, the inevitability of war was almost certain, despite both British and French politicians' apparent efforts to maintain peace at all costs against the growing threat of Nazi Germany.

As early as 1935, a government White Paper released on 3 March of that year outlined plans to increase defence spending to counter the danger of German rearmament. The main emphasis was on defence from air attacks, and further announcements made on 22 May confirmed that the Royal Air Force was to treble in size over the next two years. The following year, a new prototype aircraft, the Supermarine Spitfire, took to the skies for the first time, and on 15 July 1938, the government confirmed an order for 1,000 of these new fighters for the RAF as part of its preparations for war.

Following the Munich crisis of September 1938, fears of conflict intensified. Only a few months earlier, Winston Churchill had warned: 'We seem to be moving towards some hideous catastrophe.' Gas masks were issued and air-raid shelters were constructed. Park Works made its own preparations, training works ARP wardens and fire crews.

The realisation that war was imminent came as no surprise to David Brown & Sons. Like most firms with defence contracts, rearmament brought a welcome boost to business following the economic slump of the 1930s. For several months now, orders related to munitions production had been flowing in. As well as manufacturing tank gearboxes, the company had also been awarded an important contract to act as a shadow factory supplying aero-engine gears to Rolls-Royce. The scramble to increase the strength of the RAF led to further orders for gears which were used in the Merlin engines for the new Spitfire fighters.

It was only the beginning of what was to prove to be an extensive and varied programme of wartime production for David Brown & Sons. Against this backdrop of important military contracts, the company persevered with the

LEFT:
Women workers at David Brown's Penistone plant. During the war, the foundry employed over 200 female staff.

introduction of its new farm tractor, spurred on by an edict issued by the government on 3 May 1939 urging farmers to plough up grazing land to increase food production to counter any possible U-boat blockades. The campaign for food had begun and David Brown was ready to answer the call to mechanisation.

The David Brown Tractor

Work on designing and developing David Brown's own tractor had been under way while the Ferguson Type A was still in production, and the start of the project is believed to date back to early 1937. Initially, the plan had been to provide an 'improved' Ferguson-Brown, but it soon became evident that Harry Ferguson was unlikely to accept anything developed by the Park Works engineers. Realising that the differences between him and Ferguson were past reconciliation and that the partnership was going nowhere, Mr David Brown gave Albert Kersey the go-ahead to create a completely new model secretly – a new tractor that was to proudly bear the David Brown name.

Brown himself took a close interest in the project and personally canvassed farmers (usually while out hunting) to determine what was needed to overcome the sales resistance being experienced by the Ferguson-Brown. It was obvious that any new design had to be more powerful and stronger than the Ferguson and had to be able to handle both mounted and trailed implements. This latter point was deemed to be of upmost importance because Mr David Brown felt that sales of the Ferguson tractor had been hampered by the cost of its special implements. The Fordson tractor still dominated the market and the advertising for this basic but competitively priced machine was pertinent: 'You don't have to buy new implements for the Fordson... (it) works my existing tools, saves me money and saves the nation valuable materials.' A valid point that David Brown was keen to take on board. The next step was to investigate the opposition.

David Brown's only experience of tractor building had been influenced by Ferguson's ideas and the company wanted to know what other manufacturers were doing. A British Fordson from Dagenham and four or five leading makes of American tractors, including a two-cylinder John Deere B, an Allis-Chalmers Model U and a Massey-Harris Pacemaker, were procured and stripped down in a screened-off area on the ground floor of the Ferguson-Brown assembly shop. It was an exercise that provided valuable information and helped formulate ideas for the design of the new tractor.

At the commencement of the secret tractor project, Mr David Brown laid down several guidelines for his engineers to follow. In particular, Albert Kersey and his small team of five or six engineers were briefed to design a tractor that would be simple to service and could be easily repaired in a basic workshop anywhere in the world without the need for any special facilities.

Kersey was installed in a small back room next to the managing director's office and his work was personally supervised by both Brown himself and his technical director, Arthur Sykes. The first step, and probably the most important stage in the development of the tractor, was designing and

RIGHT:
The partly assembled engine and frame for the prototype David Brown tractor at Park Works in 1938. The cast-iron mainframe formed the lower half of the crankcase and bell-housing. Note the position of the air-cleaner.

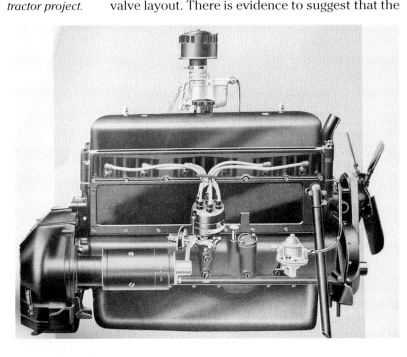

BELOW:
The six-cylinder Bedford truck engine lent many of its features to the David Brown design. The connection came through one of Vauxhall's engine designers, Alex Taub, who consulted on the tractor project.

building the engine. Kersey was keen to move away from the antiquated side-valve design inherited from Coventry Climax and favoured an overhead-valve layout. There is evidence to suggest that the Meadows engines remembered by Fred Meadowcroft, probably the overhead-valve 4EL/AV models, were considered for the new David Brown tractor but dropped for reasons of cost or reliability in favour of a new power unit designed in-house.

David Brown had learnt how to manufacture engines while building power units for the Ferguson-Brown and Kersey already had some knowledge of engine design from his time at Gilford. However, the company's experience in this field was still limited and so David Brown approached a leading engine designer, Alex Taub of Vauxhall Motors, for advice. At the time, Vauxhall, which had become part of the American General Motors Corporation in 1925, was recognised as one of the country's leading motor manufacturers. It was second only to Rolls-Royce and Daimler in terms of reputation and stature, and its design team was thought to be one of the best in the world.

Mr David Brown had already forged links with the Luton company following his exploits with the Vauxhall Villiers racing car. During his racing days, he had cultivated many friends and contacts within the motor industry, and his relationship with Vauxhall was evidently strong enough for its board to agree readily to Alex Taub acting as consultant to Albert Kersey on the tractor project.

Taub was a great proponent of lean-burn combustion. Born in London to an impoverished Jewish family, he had been sent to America at a young age to stay with relations and was educated there. He eventually joined the Chevrolet division of General Motors as an engineer and was involved in the development of the renowned six-cylinder Chevrolet engine, often dubbed 'the cast-iron wonder'. In 1936, he was seconded to Vauxhall Motors and on his return to England he designed the incredibly economical (for the time) four-cylinder engine for the Vauxhall Model 10 car that was launched in 1937.

While Albert Kersey worked on the designs for the tractor engine, Alex Taub vetted the drawings and made suggestions as the new power unit began to take shape. There was no disguising Taub's Vauxhall heritage because the tractor engine was very similar in layout to the latest Bedford truck engine, albeit in four– rather than six-cylinder form.

Vauxhall's range of Bedford W-type trucks had been launched in 1931 and were powered by an improved version of Taub's six-cylinder Chevrolet motor. The Bedford engine was a long-stroke and slow-revving unit that had already won itself an enviable reputation for reliability and great lugging power. The David Brown engine had similar dimensions, with a 3½ in. bore and a 4 in. stroke giving a capacity of 154 cu in. (2,523 cc). Both power units had overhead valves and a full-pressure lubrication system.

On the Bedford engine, the oil was circulated by a rotary gear-type pump carried in the upper half of the crankcase and driven by skew gears from the camshaft. The distributor for the electrical system was driven in tandem with the pump by the same skew gears. An almost identical arrangement was used for the David Brown engine to drive a vertical magneto. The advantage of the system was that should the oil pump fail, the ignition unit would cease to function and stop the engine before any damage was done. It was an arrangement that was fairly unique at the time because the magnetos fitted to most farm tractors were usually driven off the timing gears.

There were several other similarities between the two engines, even down to the proposed position of the starter motor. Taub also applied his 'lean-burn' thermo-dynamic principles to the cylinder head, which was designed to provide minimal fuel consumption, ensuring that David Brown tractors became and always remained the most economical machines on the market. Nothing was ever made of the Bedford influence, but Alex Taub's contribution allowed David Brown to state in its early sales literature that the tractor engine

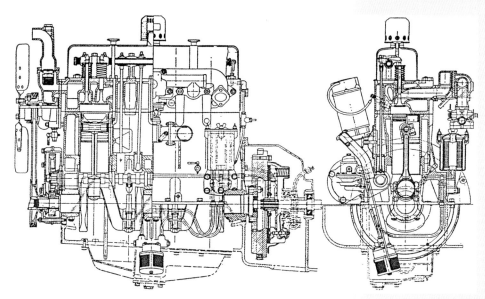

BELOW:
A sectional drawing of the production David Brown engine. It was a four-cylinder power unit with a 3½ in. bore and a 4 in. stroke. Note the vertical magneto driven in tandem with the oil pump, which was one of the features copied from the Bedford engine.

'was conceived by one of the world's leading exponents of engine design and attains an unequalled standard of efficiency and economy'.

On the down side, the Bedford engine was renowned for its thin blocks and problems with frosting due to the position of the drain tap, which allowed pockets of water to remain after the block was drained. Sadly, the David Brown design inherited similar characteristics, and cracked blocks were a feature of the early engines.

It has to be emphasised, however, that Kersey's engine was only *influenced* by the Bedford design and there remained plenty of differences. The David Brown engine also bore some similarity to the Meadows power units, particularly in the design of the timing gears as well as the use of a Solex carburettor in preference to the Zenith unit fitted to the Bedford motor.

Fred Meadowcroft strongly believes that there

was a Meadows connection and remembers the name often being bandied about in the tractor engineering department at Park Works. Like Mr David Brown, Henry Meadows started in business as a gear manufacturer. His Fallings Park Works in Wolverhampton had been producing three-speed gearboxes for cars since 1920 and was also supplying petrol engines for a variety of leading motor manufacturers, including Lagonda, Invicta, Lea Francis, Bean and Fraser-Nash. Mr David Brown had used a Meadows gearbox when he built his trials special and, with his fancy for fast cars, he would certainly have shown an interest in the company's work. During the 1930s, Meadows began building tank power units and transmissions, and some of their gearboxes were subcontracted to David Brown & Sons. However, there seems to have been no other official tie-up between the two companies and it seems likely that the Meadows engine may have been only one among several considered as possible power units for the new tractor.

One major difference between Kersey's engine and either the Bedford or Meadows power units was that it had wet sleeve liners, which were specified at Mr David Brown's insistence. The liners were renewable and eliminated the need to re-bore worn cylinders, thus simplifying the task of overhauling the engine. It was not a new concept because the American Advance-Rumely company had used renewable cylinder sleeves on its tractors as early as the 1920s, but the David Brown model must have been one of the first mass-produced tractors to be fitted with wet liners.

Another unusual feature of the David Brown engine was the circulation of the cooling system. Water was pumped to a long oval tube inserted into the cylinder head. The tube was made from sheet metal and had slots drilled into it to direct high-velocity jets of cold water on to the exhaust-valve seatings and spark-plug bosses. It was a concept that had been pioneered in America by Packard, another engineering-led company.

The pump only forced the circulation of water around the head. The block relied on a thermo-syphon system; a warm block meant lower fuel consumption and if the cylinders retained their heat then the engine would run more efficiently on vaporising oil. A vaporiser was provided to allow the engine to be started on petrol and then changed over to paraffin or vaporising oil once it had reached its correct operating temperature. Early manifolds relied on the coolant to heat the vaporiser. Ultimately, it was not that efficient, but it was still far better than the Gladwell unit fitted to the Ferguson-Brown.

Harry Ricardo, the renowned diesel engineer and a friend of David Brown's from his motor racing days, was also asked to appraise Kersey's designs. Ricardo suggested a number of small changes, including strengthening the crankshaft, which would make it easier to turn the engine into a diesel in the future.

The first prototype tractor engine was test run on a Heenan & Froude dynamometer next to the Ferguson-Brown assembly line on 13 December 1938. It developed 35 hp and had two governor settings giving speeds of 1,300 and 2,000 rpm

BELOW:
The prototype David Brown tractor engine being run for the first time on a Heenan & Froude dynamometer in the tractor assembly area at Park Works. The date is 13 December 1938 and Ferguson-Brown tractors can still be seen coming off the assembly line in the background.

The planning and design of the tractor itself ran parallel to the engine developments. As work progressed, Kersey moved his team into Park Cottage, recently vacated by Frank Brown and due for demolition to make way for further expansion. He was assisted by Alan William McCaw, whose specialities were transmissions and hydraulics. 'Bill' McCaw was only thirty-four when he was recruited by David Brown in January 1937, but he already had a wealth of experience behind him and had spent the previous eighteen years working for a number of leading engineering concerns, including most recently the British Thomson-Houston Company of Rugby.

It is impossible to stress too highly the importance of Kersey and McCaw's work. The development of this first prototype was pivotal to the future of David Brown tractors and many of the concepts embodied in it remained fundamental to the company's designs for the next fifty years.

The feature that was unique to the original David Brown tractor was the design of the mainframe. With the conventional method of farm tractor construction, as pioneered on the Fordson, copied on the Ferguson-Brown and preferred by most of the contemporary manufacturers, the engine, gearbox and rear axle were connected by vertical joint faces. The disadvantage was that for major overhauls, or even just a simple clutch repair for that matter, the tractor had to be split - never an easy task, requiring a wheeled trolley or similar and a smooth level surface. It was not a job that could be tackled lightly without special tools or be easily carried out in the field.

As Mr David Brown had decreed that his tractor should be simple to repair on the farm without the need for special workshop facilities, Albert Kersey came up with the idea of a cast-iron mainframe. This single-piece grey-iron casting formed a horizontal joint from which the self-contained power unit and transmission components could be removed without disturbing each other, thus making it unnecessary to split the tractor.

The mainframe, incorporating the footplates, was dowelled and bolted to the rear-axle housing

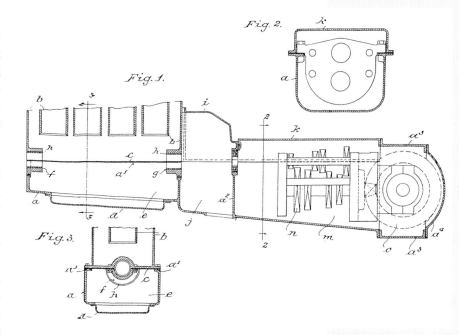

ABOVE:
Detail from the specifications of patents granted to David Brown in November 1939 covering the design of the tractor mainframe and the construction of the gearbox. It clearly shows the horizontal joint that allowed the main components to be removed for ease of servicing. Albert Kersey's name was recorded on the patents as being the inventor.

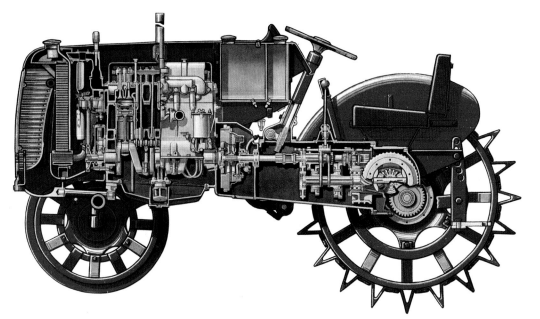

LEFT:
A sectionalised layout of the production David Brown tractor showing the main features of the engine and transmission. Many of the concepts embodied in the design remained fundamental to the company's tractors for over fifty years.

and a round front-extension casting that partly supported the front axle and radiator. It not only formed the lower half of the crankcase and bell-housing, but also accommodated the gearbox and held the transmission lubricant. It was a simple procedure to lift out either the engine or gearbox with a block and tackle by unbolting at the joint face, after first removing the gearbox cover, while still leaving the tractor standing on all four wheels. The clutch could also be removed without disturbing the engine or gearbox.

Naturally, with David Brown's gear heritage, the gearbox itself was a masterpiece of engineering. It was a robust four-speed and reverse unit, constructed as a compact, self-contained unit incorporating the spiral-bevel wheel and differential. It was formed from an open frame of two cast-iron endplates and four steel spacer-bars. The endplates housed the bearings, which in turn carried the shafts and gears.

The rear axle supported the final-drive reduction units. These were also designed so that they could easily be unbolted and removed by one man after placing a jack under the rear axle and removing a rear wheel. The whole concept, including the wet cylinder liners, came together to provide a tractor that was simplicity itself to service.

After David Brown Tractors Ltd was re-formed in February 1939, Kersey was named as chief engineer. He was tall with jet-black hair and has been described as a pleasant but quiet and unassuming man who was never keen to assert his authority. McCaw was chief draughtsman, but it is unclear as to who exactly was in charge of the design and who was responsible for what. For a time, it seems as if both men were doing the same job. Charles Hull, who joined the company as a design draughtsman in 1940, worked under Albert Kersey for a considerable time and remembers him as 'a modest man and a gentleman of the old school'. He believes Kersey's contribution to the development of the David Brown tractor has often been undervalued and not always given the recognition it deserved.

Certainly, it was Kersey's name that was recorded as the inventor on the British patents granted in November 1939 covering both the design of the tractor mainframe and the arrangement and construction of the gearbox. Both features were a radical departure from conventional tractor design and remained intrinsic to virtually every David Brown model ever built.

Bill McCaw collaborated with Kersey on the gearbox, but had really been recruited for his experience in hydraulics. Mr David Brown planned to launch the new tractor with a hydraulic lift so that it could offer the same advantages as the Ferguson-Brown and use similar mounted implements. However, he wanted the hydraulic unit to be a self-contained 'bolt-on' accessory. The farmer then had the option of buying the tractor without hydraulics to tow his existing trailed machinery and then adding the hydraulic unit at a later date if he so wished. The idea was to spread the cost by gradually changing over to mounted equipment.

The power lift bolted onto the rear axle in place of the differential cover. The original prototype unit was not a very compact adaptation. The ramshaft was mounted further out from the back of the tractor than the top-link position, with the lift arms extending back towards the rear axle. The unit was also combined with a power take-off shaft.

Powered by the same type of four-cylinder Dowty pump as was fitted to the Ferguson-Brown, the lift had two ram-cylinders and incorporated some sort of draft-control system with top-link sensing. The top-link was connected via a fork to a sensing mechanism mounted on top of the unit's casing. The adoption of draft control would appear have been a foolhardy move, but at that time David Brown's engineers were not yet fully aware of just how many features of the Type A tractor were covered by Ferguson-owned patents.

The first prototype tractor was assembled in a quiet corner of Park Works, well away from prying eyes. No one seems exactly sure as to where this happened because only Kersey's small team and Mr David Brown himself were privy to that information. Some work was carried out behind the screen on the ground floor of the tractor assembly shop in the concrete building. The screened-off area was marked 'private' and referred to as 'the forbidden territory', but this may have been a bluff to antagonise Harry

Ferguson during his increasingly infrequent visits to the production line.

The design of the prototype was finalised after Kersey's team returned from their Christmas break in late December 1938 and the new engine was fitted into the tractor. Work took on greater urgency in view of the pending split with Ferguson and Mr David Brown's insistence that his new tractor had to be ready for the following July to be launched at the Royal Show, the shop-window of the farming community. He was negotiating the purchase of new premises to be dedicated to tractor manufacturing and was keen to get production under way before the company was swamped with armament orders.

During the first week of 1939, the prototype tractor, registered BVH 10, was ready for road and field trials. Ernest Kenyon, Cyril Marshall and Bill Harrison were recruited to Kersey's team to act as test drivers. The project was nearing completion, but niggling problems led to much burning of the midnight oil and kept the tractor from its planned inaugural run.

LEFT:
The prototype tractor under construction with the power lift in place. The prototype hydraulic unit had two ram cylinders. It incorporated a power take-off as well as some sort of top-link sensing for draft control. Note the arrangement of the lift arms.

BELOW:
Cyril Marshall at Park Works with BVH 10, the original prototype David Brown tractor, which ran for the first time in late December 1938. Note how much ground clearance the tractor has. This was later reduced to give the machine greater stability.

Rumours that the tractor was finally ready percolated around the plant one morning in early January after several of the experimental team failed to put in an appearance after the previous night's work had dragged on into the early hours of the morning. As details of their endeavours began to emerge, it appeared that the tractor had indeed made its first run and in spectacular fashion!

The story of that night's exploits became almost legend within the tractor division. It seems that after a particularly long and unfruitful session trying to sort out the prototype, Mr David Brown told his engineering team to call it a day before he himself left for his new home, Durker Roods at Meltham. Believing that they were close to cracking the problem, the team decided to carry on for a few more hours, eventually managing to get the tractor to run before retiring to a local hostelry to celebrate.

After leaving the public house, the three test drivers, Kenyon, Marshall and Harrison, no doubt fortified by the local brew, decided to take the tractor up to Mr David Brown's home and surprise him with the news that it was running. This was no mean feat in itself: Meltham is a good four miles from Park Works and the road is narrow, twisty and steep. Negotiating it on an untried tractor without lights in the middle of a cold winter's night must have been an unforgettable experience! Brown was evidently raised from his bed by the bark of the straight-through exhaust at two in the morning.

Rather than being disgruntled at being disturbed in the early hours, Brown was overjoyed to hear the tractor running and emerged from Durker Roods in his dressing gown with a whisky bottle in his hand to toast the arrival of his new 'baby'. He was also keen to drive the tractor and took several turns around the grounds before discussions turned to more mundane matters, such as what colour the machine should be painted. While everyone voiced their opinion, Brown rushed back inside the house to fetch his scarlet hunting coat from a peg in the hall. He threw the coat across the tractor bonnet with the challenge that the paint should match the jacket.

Mr David Brown had always shown more interest in the appearance of the tractor than its mechanical development. He wanted it to have distinctive looks with modern and stylish lines. The bonnet had to be free of all excrescences and driver comfort had to be taken into consideration. While Albert Kersey dealt with the engineering concerns, Brown evidently experimented in his office with different styling concepts by making models of the tractor out of Plasticine.

A wooden styling mock-up of the tractor had been built during 1938 on a Ferguson-Brown chassis in the paint section on the top floor of the concrete building. Hidden by screens, it had been used to try out a number of different paint schemes, including Ferguson's battleship grey, bright blue and a vivid red. Red became the preferred colour and, at Mr David Brown's insistence, the shade chosen was the same hunting pink as the uniform of his local Badsworth hunt.

The first prototype tractor, BVH 10, was painted grey, partly because it had been built before the colour of the new livery had been finalised and partly to disguise the tractor on test so that from a distance it looked like a Ferguson Brown. The initial test programme was both extensive and exhaustive, not least for the drivers Kenyon, Marshall and Harrison, who took turns at the wheel to keep the tractor running almost continually night and day.

Most of the field trials were carried out on Crosland Moor above Lockwood. Mr David Brown owned land on the moor where he had established gallops for his horses. Some of the land was already used for field instruction of the Ferguson Brown tractor, and a shed had been built to house the implements for the training school run by Fred Meadowcroft. Part of the shed was sectioned off for use by the experimental engineers and remained out of bounds to the students and instructors for the duration of the prototype testing.

The site on Crosland Moor was sufficiently isolated to allow testing to be carried out in secrecy during the day. At night, the prototype tractor took to the lanes around Huddersfield and Meltham for road testing. One of the most popular routes was the three-mile climb out of Meltham across Meltham Moor to the Isle of Skye inn on the Holmfirth to Manchester road before

LEFT:
The prototype David Brown tractor underwent extensive field trials on Crosland Moor during the early months of 1939. Test driver Bill Harrison is at the wheel.

returning via the Ford Inn, crossing Thick Hollins Moor and back to Meltham. Bearing in mind that most of the testing took place in the first quarter of 1939, it would have been a severe trial of both men and machine with the sleet and snow driving in across the moors.

Although the testing of the new tractor was supposed to be a closely guarded secret, it is interesting to note that a poem recording the trials and tribulations of the night-time runs appeared in the March 1939 edition of the company magazine 'Contact'. There was even a cartoon of a tractor (albeit a Ferguson-Brown model) stuck in a snowdrift. The poem began:

'Twas a terrible night on the moors sir, with snow six feet deep on the ground,

LEFT:
A wooden mock-up built at Park Works was used to finalise the styling and colour scheme of the new David Brown tractor.

RIGHT:
Walter Hill, the general manager of David Brown Tractors Ltd, tries the prototype machine out on Crosland Moor. The test programme was exhaustive with the tractor working on the moor during the day and running on the road at night. Note the forward position of the exhaust silencer.

> And save for the chug of the Tractor there wasn't the ghost of a sound.
> 'Twere a night fit for no man nor beast sir, but still there was work to be done,
> For our tractor was out on the night shift – she were due for her very last run...'

For the initial field trials, the tractor was not fitted with the power lift and was used with trailing implements including a two-furrow Oliver plough. The latter implement had been procured from the USA and was delivered to Park Works in component form in a crate. Fred Meadowcroft was given the job of assembling it before taking it out on to Crosland Moor for evaluation.

Tests were carried out on both pneumatic tyres and steel wheels. As the programme

RIGHT:
The prototype David Brown tractor was evaluated with Ferguson mounted implements including this two-furrow plough. The trials highlighted several failings in the design, particularly with the power lift which proved to be unreliable.

progressed, tests were conducted with the power lift and the tractor was evaluated with the range of Ferguson mounted implements kept at the training school shed. The trials were not without incident and highlighted several failings in the design, not least with the power lift unit itself, which did not perform as well as expected.

The prototype tractor was designed for rowcrop work and had very good ground clearance, but was found to be unstable on test because its centre of gravity was too high. To rectify this, the front axle, which consisted of a longitudinal tube, was redesigned with shortened swivel pins. Smaller diameter 9.00 x 24 rear wheels and tyres were substituted for the 9.00 x 28 equipment originally fitted and alterations were made to improve the steering system. The centrally-mounted steering column was also altered to give the tractor a unique offset driving position to aid the driver's vision.

In order to conform to Mr David Brown's ideals of clean, flowing lines, the Burgess air-cleaner was bolted to the side of the engine and hidden away beneath the bonnet. However, it soon became evident that for farm work the air inlet needed to be higher to minimise the dust intake. Consequently, the air-cleaner was eventually moved outside of the frame and fitted with an extension pipe to a domed pre-cleaner.

For reasons of aesthetics, the exhaust pipe fitted to the prototype tractor was mounted forward of the bonnet line, projecting vertically from the manifold in line with No.1 cylinder. This gave the tractor a very loud exhaust note and obstructed the driver's vision. The proximity of the exhaust elbow to the radiator also led to some overheating problems. For the pre-production tractors, the exhaust was moved back to a point between No.2 and No.3 cylinders, which was a great improvement and meant that the engine was much quieter.

While sorting out problems with carburation, it became apparent that the driver needed to keep his eye on the water temperature when making

LEFT:
A number of changes were made to the prototype David Brown tractor during trials, including modifications to the front axle and driving position. The exhaust system was later altered as it proved to be too noisy. The tractor is on test at Meltham in a field at the back of Durker Roods with Bill Harrison at the wheel and his son along for the ride.

The David Brown Tractor Story

RIGHT:
A Land Army girl poses with one of the pre-production tractors outside the concrete building at Park Works. The original tinwork was believed to have been fabricated by a team of panel beaters from Mulliners of Birmingham. The grille on the pre-production tractors was made from sheet steel.

BELOW:
One of the pre-production David Brown tractors on trial with an Oliver plough in 1939. The tractor has the new styling with the horseshoe-shaped scuttle, and the repositioned exhaust and air-cleaner.

the changeover to vaporising oil. A calorimeter was fitted to the water outlet pipe from the engine and allowed to protrude through the radiator. Mr David Brown's desire to keep the bonnet free of all excrescences was becoming impractical, but the tractor still became one of the most stylish machines on the market.

The original prototype tractor was only fitted with basic tinwork, but while its mechanical trials were under way, plans were made to finalise the styling and design of the sheet metalwork. It is believed that the original tinwork for the pre-production tractors was fabricated by a team of panel beaters from the famous coachbuilders, Mulliners Ltd of Birmingham.

The stylish radiator grille, which was reminiscent of American automotive designs of the period, was apparently made from sheet steel for the first pre-production tractors. However because of its intricate design, the production version was constructed from four pieces of cast iron to simplify the manufacturing process. To help protect the driver from the elements (no doubt prompted by the mid-winter excursions across Meltham Moor) and to direct heat back from the engine, a horseshoe-shaped scuttle was fitted over the bonnet and extended down to the footplates. A bench seat was also provided.

A total of five prototype or pre-production models were built. Three were used for trials while the other two were prepared for show purposes. The first show and demonstration models were finished in hunting pink but had different colour wheel centres. It has been suggested that these could have been pale blue, pale yellow or even silver.

The first demonstration

LEFT:
A total of five pre-production David Brown tractors were built. This model is seen on trial with the prototype hydraulic lift and a Ferguson ridger. Problems with the early two-cylinder power lift and concerns about infringing Ferguson patents meant that the design was eventually shelved before the tractor went into production.

The David Brown tractor was held in early June 1939. Only selected members of the agricultural press were invited as a taster to the new model's public unveiling at the following month's Royal Show. Two tractors were put through their paces. One, fitted with steel wheels and an extension drawbar but no hydraulic lift, was shown working with trailed implements; while a second machine, equipped with pneumatic tyres and the prototype hydraulic unit, was demonstrated with a spring-tine cultivator.

Even though the war clouds were gathering, that year's Royal Show was a lavish affair. Opening at Windsor on 4 July, it was the Royal Agricultural Society of England's centenary show and the machinery manufacturers pulled out all the stops, filling 430 stands with the latest agricultural equipment. The *Implement and Machinery Review* magazine reported 'New tractors ... from totally unexpected sources and of unusual design'. The David Brown tractor came into this category and certainly created some interest.

David Brown Tractors Ltd had stand No.86, not far from the Royal Café in the heart of the 'Machinery in Motion' section. The stand was

LEFT:
An artist's impression of the styling for the new David Brown tractor. This picture was released to the press in June 1939.

LEFT:
The first brochure for the original David Brown tractor as launched at the Royal Show in July 1939.

The David Brown Tractor Story

RIGHT:
The new tractors were seen in the company of Ferguson-Brown models on David Brown's stand at the 1939 Royal Show, which opened at Windsor on 4 July. The new model created much interested and allegedly elicited over 3,000 orders.

busy for all the five days of the show as visitors marvelled at the modern lines of the new tractors in their lustrous hunting pink paint. Drawing the crowds, they shone like beacons next to the dull grey of the Ferguson-Brown models.

Ferguson's Type A tractor may have been at the forefront of technological development when it had been released just three years earlier, but now its functional and angular appearance looked dated when compared to what Albert Kersey and his team had produced. Harry Ferguson's reaction to the new model is not recorded, but rumour has it that he was evidently not pleased by what he saw.

The press heralded the David Brown tractor as 'an entirely new all-British machine'. Rather than concentrate on its mechanical features reports focused on its looks and much was made of its 'sleek external appearance' and 'graceful lines'. Mr David Brown was not wrong when he had insisted in putting as much emphasis on the tractor's aesthetic design as on its mechanical components.

The company claimed to have taken over 3,000

RIGHT:
An early David Brown tractor restored in the style of the machines exhibited at the 1939 Royal Show. Its hunting pink livery was chosen by Mr David Brown to match that of the Badsworth hunt.

orders for the new tractor at Windsor. The validity of these figures has often been questioned and it has been suggested that 3,000 enquiries was the more likely scenario. Had David Brown taken a leaf out of Harry Ferguson's book and been a little economical with the truth?

While it does seem doubtful that so many farmers would order a completely new tractor just on the strength of its Royal Show appearance, recent history has shown us that startling and revolutionary new models have attracted unprecedented numbers of orders at their launch (such as the appearance of the revolutionary Citroen DS car at the 1955 Paris motor show) and have gone on to become icons of their age.

The David Brown tractor was certainly revolutionary and the impact it made at Windsor was undeniable. Customers were drawn to its modern appearance and it was dubbed 'the Rolls-Royce of tractors'. Such a demand does not seem that excessive when one considers that British farmers would buy over 150,000 new tractors in the coming six years. It does not seem unreasonable to suggest that a good proportion of these would have been David Brown models if their eventual production had not been restricted by military contracts.

The new tractor was shown at Windsor with the prototype two-cylinder lift, but it was becoming increasingly evident that the hydraulic unit would have to be shelved before production began because of reliability problems and patent difficulties. Announcements were made that the tractor would go into production 'at an early date' and that demonstrations would be given in various parts of the country.

Initially, no mention was made of the fact that that tractor would be built at the new Meltham factory. This was partly because negotiations for the new site had not been finalised and partly because some of the works space had already been earmarked for defence contracts and it seemed wise not to mention its exact whereabouts during a time of national emergency.

Following the Windsor show, detailed production and sales forecasts for the new tractor were prepared and actually provided for 3,000 tractors to be built in the year ending 31 December 1940. A proposed retail price for the tractor on steel wheels was fixed at £146 6s, which it was calculated would give the company a profit of £28 6s on each machine. A range of implements was also planned and costed, but the intervention of war distorted both production forecasts and pricing structures. On 3 September, Prime Minister Neville Chamberlain declared that Britain was at war with Germany. The announcement had an immediate effect on David Brown's plans; demonstrations were curtailed and production was delayed.

To manufacture tractors, you needed steel. The Ministry of Supply controlled allocations of raw materials and the British government felt that the fledgling tractor firm should divert all its energies into wartime production in the same way that Park Works and the other David Brown factories were gearing up for munitions work. Mr David Brown agreed up to a point, but argued that tractors still had a place in the war effort. The Ministry of Supply could see some logic in this, and modified its demands to allow limited tractor production to be carried out alongside vital war work at the new Meltham Mills factory.

Meltham Mills

Meltham lies in a triangle of wild Pennine country bounded by Huddersfield, Manchester and Sheffield. It is situated to the south-west of Huddersfield, its nearest neighbour, and is less than six miles away. Nestling on the edge of the Peak District National Park, it is a sizeable community that was largely a product of the industrial revolution.

There were several mills in the area, the most extensive being the Meltham Mills complex on the western edge of the village, just off the Huddersfield road. The complex became almost a little community in its own right, joining with the hamlets of Helme and Wilshaw to form part of the larger township of Meltham.

The site of the factory is marked by a collection of buildings bounded by Meltham Mills Road and Knowle Lane. The latter climbs steeply to Knowl Top, while Acre Lane, at the southern extremity of the site, leads out on to Thick Hollins Moor. Today, the plant is only a shadow of its former self, but enough imposing structures remain to bear testimony to the existence of the home of David Brown tractor production for nearly fifty years.

RIGHT:
Meltham Mills nestles on the slopes of a wooded valley on the edge of the Peak District National Park. The photograph was taken in the mid-1980s when the plant had expanded to its fullest extent and gives some indication of its rural location.

The Meltham Mills site can only be described as dramatic - monumental buildings perched on the lower slopes of a wooded valley in the foothills of the Pennines. As the name implies, more than one mill complex resided on the site. The original Meltham Mill was a small corn mill built in 1760 by Nathaniel Dyson. The site was chosen to take advantage of a brook, the Royd Edge Clough, which ran down the valley so that the mill could be powered by a waterwheel.

The textile industry came to the valley in the latter part of the eighteenth century when William Brook established a number of woollen mills on the site. One of the earliest was the Clock Mill on Knowle Lane, which derived its name from the turret clock - a mechanical clock driven by a falling weight - that was fitted high up in the building.

It was Brook's son, Jonas, who was responsible for the real expansion of Meltham Mills. Trading in partnership with his brothers as Jonas Brook & Bros. Ltd, he moved into the manufacture of sewing cotton and erected spinning, thread and bobbin mills on the site. The secluded valley was the ideal situation for a cotton mill because it offered the moist atmosphere necessary for many of the spinning and weaving operations. Reservoirs were built to provide a plentiful supply of soft water for bleaching and dyeing and to power the waterwheels that drove the mills. Further expansion saw the family buy out other mill owners and erect more buildings until Meltham Mills covered some thirteen acres. The Brook family residence, Meltham Hall, was built near to the mills in 1841.

During the nineteenth century, the Brook brothers and their descendants became great benefactors to the area. The family was recognised for the concern it showed for the welfare of its employees, providing its workers with stone cottages, a convalescent home, a community hall, a park and even a cricket pitch all within the close vicinity of the mills. The firm also donated the money for a church and school to be built opposite the mills, and financed a railway line from Huddersfield to Meltham. The branch line, completed in 1869, terminated in the centre of the town in a hollow known as Scarr Bottom. The marshalling yard at Scarr Bottom was laid out in the shadow of an imposing Victorian worsted mill - a five-storey stone building owned by the Meltham Spinning Company.

ABOVE:
An early nineteenth century artist's impression of Meltham Mills showing the Clock Mill on Knowle Lane. This woollen mill was one of the earliest buildings on the site and probably dated back to 1760.

LEFT:
Meltham Mills plant soon after being taken over by David Brown Tractors Ltd. The site had been used for cotton production and many of the buildings were former spinning or bobbin mills, some dating back to the eighteenth century.

In 1896, Jonas Brook & Bros. Ltd was taken over by J & P Coats Ltd of Paisley in Scotland, whose wooden cotton-reels were once seen in almost all haberdashery stores. Coats later formed the Meltham subsidiary into United Thread Mills. Meltham Hall remained in the hands of the Brook family until 1944, when it was bequeathed to Meltham Urban District Council.

The economic depression of the 1930s brought about a rationalisation of the cotton industry and plans were eventually made to transfer the United Thread Mills operation to Paisley. The company ceased trading at Meltham in 1936 and the next three years were spent winding down the operation. The company was well aware of the impact that the mill closure would have on Meltham town because there was no other employment in the area. Some of the staff had been relocated to Scotland, but many faced the prospect of no jobs. Coats held out in the hope of finding a buyer to take on the factory complex and provide some sort of employment for the workers, which is when David Brown came to the rescue.

LEFT:
Meltham Hall, built in 1841, was home to the Brook family who had established the cotton mills in Meltham during the latter part of the eighteenth century. The residence was bequeathed to Meltham Urban District Council in 1944.

LEFT:
A consignment of David Brown tractors prepares to leave Scarr Bottom during the Second World War. The line from Huddersfield to Meltham was financed by the Brook family and completed in 1869. The 0-8-0 locomotive is a former NWR Class G2A, originally built at Crewe in 1898.

RIGHT:
Mr David Brown's family residence, Durker Roods, not far from the Meltham Mills plant. The house, originally built in 1870, is now a hotel.

Mr David Brown had lived in Meltham since 1927 when he bought a family residence in the town. His house had originally been built in 1870 by a local businessman, Arthur Carlow Armitage. It was called Durker Roods, which in Anglo-Saxon means 'house on the marsh'. Charles Brook JP, a descendent of the Brook family, had lived there from 1901 to 1917.

The house also had extensive stabling for Mr David Brown's horses. Apart from riding with the local hunt, Brown had taken a fancy to racehorses, buying his first at a sale in Dublin in 1933. Within a short time, he was recognised as an upcoming owner-trainer with several successes to his name. A small training quarters was established in the grounds of Durker Roods, which was literally only five minutes walk from Meltham Mills.

Brown was well aware of the potential offered by the mill complex, and his initial investigations showed that the buildings could easily be adapted to modern line production and offered ample scope for accommodating the necessary feeder-lines. However, it was a big site with over 400,000 square feet of usable floor-space, and it was not until his tractor project came to fruition in 1939 that he felt he could justify the purchase of such a large property. Even then, he still waited to see what sort of reception his new tractor would get at the Royal Show before beginning negotiations in earnest. The deal was concluded during August and Coats were happy to sell the site for the favourable price of £14,000 provided that the tractor company would guarantee employment for at least half of the 800-strong workforce. An official announcement confirming the take-over was made on 15 September.

In many ways it was a repeat of the Penistone take-over; the deal benefited both the company and the local community and David Brown got a ready-made factory complete with an abundance of local labour. Admittedly, they were textile workers, but they were semi-skilled men and women who were adaptable and receptive to retraining into the ways of engineering practice.

The arrangement with United Thread Mills was for the David Brown plant and machine tools to be gradually introduced over six months to allow time to train and absorb the local labour. The cotton and spinning machinery would be cleared out over the same period. During the phased changeover, Coats let Brown lease the premises for a nominal rent with the option to purchase once the move was complete.

A training scheme for the former textile workers was set up at Park Works in the concrete building.

RIGHT:
Meltham Mills plant during the Second World War. The former weaving sheds in the foreground housed the tractor assembly while the five-storey building in the background was used for aero-gear production.

LEFT:
The machine shops for the Meltham tractor plant from the junction of Mill Bank Road with Meltham Mills Road. The chimney was for the boiler house, which had been adapted to provide heating for the factory buildings.

LEFT:
Limited tractor production began at Meltham in 1939 with the assembly line established in the former United Thread Mills machine shops. Initially less than 100 people were involved in the tractor manufacturing operation.

LEFT:
David Brown tractors outside the remains of the old Clock Mill in Knowle Lane in 1946. The building was eventually demolished in 1959.

Those Park Works employees who had been involved in tractor manufacture were given the choice of remaining at Lockwood and being absorbed into the parent company or transferring to Meltham.

David Brown personnel began to infiltrate Meltham Mills during September. Len Craven was one of the first to arrive and as one of the storekeepers was given the responsibility for setting up a goods-receiving section. The machine tools moved in as the cotton-spinning machinery moved out. The tractor assembly line was established in an old weaving shed where United Thread Mills continued to operate its own machine shops at the far end of the line until the end of 1939.

For those David Brown men used to the purpose-built workshops and organised layout of the heavy-machine shops of Park Works, it came as something of a culture shock

RIGHT:
Although taken in the 1950s after a few additions had been made to the plant, this photograph of the Meltham Mills complex gives a good idea of the layout of the site. The tractor assembly shops are in the centre with the buildings for tank gearbox production to the right on the opposite side of the factory service road. In the foreground are the machine shops on the corner where Meltham Mills Road joins Knowle Lane, which comes in from the right past the main offices. The Aero Block is on the extreme left of the photograph.

RIGHT:
David Brown's contribution to the war effort was enormous and all its factories had vital defence contracts. Production at Park Works included manufacturing all types of marine propulsion units. This gear unit for a destroyer was capable of delivering 20,000 hp.

when they arrived at Meltham. They were faced with a hotchpotch of buildings, some over 150 years old with leaking roofs and weakened floors. The mills were built on different levels and had a myriad of floors connected by spiral staircases. The site was a maze of dark, narrow alleyways and subterranean passages. Even one of the waterwheels was still in place.

It was immediately obvious that there would be problems associated with the layout of the site for volume tractor production. Future redevelopment or expansion would also be limited because there was nearly a 200-ft drop between the buildings on the highest and lowest levels. However, once work in the plant got underway, it was soon found that the most efficient method of ensuring continuous production flow was by moving components from the feeder-lines to the main assembly lines by using tug-tractors.

One by one, the rooms were cleared of silken thread, cotton and ribbon to make room for milling, grinding and boring tools, broaching and gear-cutting machines, heat-treatment and engine-test equipment. Limited production began towards the end of 1939 with a staff of less than 100 people. A new era in the history of Meltham Mills was ushered in with its transformation into the headquarters of David Brown Tractors Ltd.

The Aero Block

Once war broke out, almost all the David Brown group's considerable resources were directed towards the war effort. The organisation came under the direction of the War Office and was swamped with orders from the Ministry of Supply, the Admiralty, the Ministry of Aircraft Production and the Ministry of Home Security. All the plants

were affected and to quantify the full extent of David Brown's wartime production is almost impossible.

Like the Windmill Theatre, Park Works never closed and its machine shops were running at full stretch right through the day and night for nearly six years, turning out gearboxes, drive gears and ring gears for everything from military vehicles and tanks to barrage-balloon winches, wind tunnels and searchlights. On the marine side, the plant produced, among other things, the main-propulsion gear-units for all types of vessels, catapults for aircraft carriers and the timing gears for submarine diesel engines. More works buildings were erected and even Park Cottage was eventually demolished to make way for further expansion.

Penistone's contribution to the war effort was considerable: the foundry produced nearly 20 percent of the total UK output of bullet-proof armoured plating as well as the steel casing for the deep-penetration 12,000 lb 'Tallboy' bomb that was first dropped on the Saumur tunnel in Germany in 1944. Although not well publicised, there is evidence to suggest that the plant was also involved in the development of both the 22,000 lb 'Grand Slam' bomb and Barnes Wallis's bouncing bombs that breached the Ruhr dams. From 1942, Penistone began producing specialised castings for aircraft parts. Smaller aerial bombs and tank drive-sprockets came out of the Salford foundry belonging to the P. R. Jackson subsidiary.

In order to increase David Brown's war production capacity, the organisation acquired the London-based firm of C. J. Fitzpatrick in 1940. The company, which was later renamed David Brown Gears (London) Ltd, specialised in marine and aircraft parts. During the war, it produced both naval propulsion units and aero-engine components. As an indication of just how important the defence contracts were to the organisation, David Brown & Sons' annual turnover, which was hovering just below the £1.5 million mark in 1939, had rocketed to over £3.1 million by 1941.

Aero-gears became a very crucial part of David Brown's wartime work and were essential to keep aircraft production flowing during the Battle of Britain. David Brown's original contract, negotiated with Rolls-Royce by Allan Avison, was for the supply of gears for Merlin III engines as fitted to the Fairey Battle light bombers. This work was later extended to include Merlin engines for the new Spitfire fighters.

As the work intensified during the early part

LEFT:
David Brown had been involved in munitions work for the aero industry since 1938 and Penistone began producing specialised castings for aircraft landing-wheel brackets in 1942.

LEFT:
The Aero Block from Meltham Mills Road. David Brown's aero-gear division was relocated here in 1940. The five-storey building was a former cotton thread mill built in 1928.

ABOVE:
The aero-gears produced at Meltham were the ultimate in precision engineering. Components made for the Rolls-Royce Merlin included engine, reduction and supercharger gears.

of 1940, the aero-gear division quickly outgrew the space allocated to it at Park Works and was relocated to Meltham. The division was housed in one of the newest buildings on the Meltham Mills site, a former cotton thread mill that was built in 1928. It was a five-storey red-brick structure that became known as the 'Aero Block' – a name that has stuck to this day even though the building was officially 'C-Block'. It was located next to a dam that had formed a reservoir to feed one of the steam engines that had originally powered the mills, the boiler of which was adapted to provide heating for the mill complex.

At the height of the Battle of Britain, all the other factories producing gears for the Merlin engines had been put out of action by German bomber raids. Meltham Mills escaped unscathed largely due to its secluded position in the wooded valley and its isolated location away from any large conurbation. However, no chances were taken; the Aero Block was camouflaged with green and brown paint, and cardboard cut-outs of cows were fixed to the roof in an attempt to make it look inconspicuous from the air. As a further precaution, Hispano cannons were mounted on the roof and manned night and day by the factory's home guard unit as defence against any curious enemy aircraft. Incidentally, clusters of lights were laid out on one of the nearby moors to act as a decoy for Mancester.

Through 1940 and 1941, the gears produced in the Aero Block became vital to national security. As the company was to later claim: 'Ours was the only factory producing gears for the Spitfire when that fighter stood between us and destruction'. Actually, for a time Meltham was also the only plant making gears for the Spitfire's partner aircraft, the Hawker Hurricane. It appears that German military intelligence was unquestionably aware of this last bastion of aero-gear production in England, and even knew that it was an engineering firm in the north of England but was never able to pinpoint its exact location.

Legend has it that William Joyce, better known as Lord Haw-Haw, mentioned Meltham Mills in all but name in one of his German propaganda broadcasts and suggested that it would be bombed shortly. Luckily for Meltham, it turned out to be an idle threat. The closest the Luftwaffe got was when it mistakenly bombed a nearby chicken farm at Wilshaw. Either they believed the farm was Meltham Mills or (more likely) it was a stray bomb dropped by an aircraft returning from a raid on Manchester. The first the local population knew of the bombing was when they

RIGHT:
Propeller reduction-gears from Meltham kept the Supermarine Spitfire fighter in the air during the Battle of Britain. At this time David Brown had the only factory still producing these types of aero-gears and its work was of vital national importance.

awoke in the morning to find the area carpeted with feathers as far as Holmfirth four miles away! Actually, the only disruption to production was caused by nature when heavy overnight rain led to severe flooding of part of the works in 1942.

During his eleven-month tenure in charge of the Ministry of Aircraft Production from 1940 to 1941, the newspaper magnate, Lord Beaverbrook, was in regular contact with Meltham, continually praising the company's efforts while at the same time pushing for increased production. Mr David Brown took great pleasure in circulating a story that the minister had visited a leading aircraft firm and insisted that its latest model be up in the air and ready to bomb Berlin in two days. The story went that the firm reorganised its factory schedule to get the plane ready on time. Within 48 hours, the aircraft was over Berlin, but when the aimer opened the bomb doors, out fell two members of the factory's night shift!

Joking apart, Beaverbrook's pleas did not go unheeded. The workforce in the aero-gear division stepped up production and worked eighteen-hour shifts in an attempt to keep the Royal Air Force in the skies. As a gesture of goodwill, Mr David Brown presented twenty sets of gears gratis to the country. His employees followed suit, voluntarily working without pay to produce another ten sets, thus equipping thirty Spitfires for the war effort. With all the firms involved in the aircraft industry 'doing their bit', fighter production actually exceeded estimates and the RAF never had less than 100 Spitfires or Hurricanes in reserve.

In January 1941, the David Brown organisation was employing some 5,700 people. Park Works had a staff of 3,000 while the workforce at Meltham had risen to 950. Of these, 300 worked in the Aero Block.

The staff in the aero-gear division consisted of both men and women, mainly former textile workers drawn from Meltham and the surrounding villages. The vital gears that they were producing for the Merlin were actually the propeller reduction gears. Housed in casings at the front of the power-plant, they reduced the V-12 engine's revolutions down to speeds that could be transmitted safely to the propeller. Usually the propeller turned at just under half the speed of the engine.

Some of the Merlin variants were putting out in excess of 1,700 hp and the reduction gears had to withstand the strain of this power. They were the ultimate in precision engineering. Although they were large in diameter, their finely machined walls were only ⅛in. thick. The gears were cut and splined by machine and then dressed by hand.

Dressing the gears involved giving them a highly polished finish with no detectable file marks. The finishing was carried out by the female staff, known as dressers, using emery cloths for the final polish. Fred Meadowcroft's wife, Connie, who was one of the ladies in the Aero Block, remembers her work with pride: 'When we had finished with the gears, they were beautiful and shone like silk'.

David Brown gears were fitted to several versions of the Merlin as used in the Spitfire, Hawker Hurricane, Avro Lancaster, Boulton Paul Defiant and de Havilland Mosquito aircraft. Other Rolls-Royce engines using reduction gears made at Meltham included the Vulture, Peregrine and the 2,050 hp Griffon power-plants. On top of this, the Aero Block produced timing gears for the Bristol Centaurus and Hercules radials (as fitted to the Handley Page Halifax and Short Stirling

ABOVE:
An indication of the importance of Meltham to the RAF during the Second World War: the Merlin engine powering this Hawker Hurricane MkII fighter bomber relied on gears from the Aero Block while a David Brown towing tractor was used for 'bombing up'.

ABOVE:
Women workers in the Aero Block at Meltham dressing aero-gears. After the components were filed smooth, they were polished with emery cloths until no file marks remained.

RIGHT:
Several different types of aircraft used aero-gears manufactured at Meltham, including the de Havilland Mosquito that was powered by two Rolls-Royce Merlins. A David Brown towing tractor hauls the bomb load.

bombers), gears for Napier Sabre engines and supercharger gears for some of the later Spitfire Merlins. A manufacturing line for Dowty hydraulic pumps, which were used for raising and lowering the aircraft undercarriage, was also established in the building.

The vital importance of the aero-gear work at Meltham to the war effort was finally revealed in September 1944 when Albert Carr, an assistant to the United States' War Production Board, published a report entitled 'The Five Big Blunders that lost the Axis the War'. Alongside strategic errors, such as Hitler's attack on Russia and the effect of Pearl Harbour in bringing America into the war, Carr cited the Luftwaffe's failure to bomb a 'single Yorkshire engineering works (that) was turning out virtually all the heavy gears for the British plane industry' as one of Germany's greatest mistakes.

Towards the end of the war, Meltham Mills and a nearby subsidiary operation at Scarr Bottom

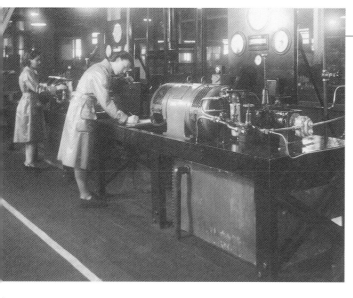

were employing in the region of 2,500 people, including over 1,000 women workers. However, by the time VE Day arrived on 8 May 1945, the work of the aero-gear division was over. Since 1940, it had produced over half a million sets of gears for the Spitfire and other types of aircraft, as well as in the region of 6,000 undercarriage hydraulic units. It was not the end of David Brown's connection with the aeronautical industry because aircraft component production continued at Penistone. Neither was it Meltham's only contribution to the war effort because the work of its tank gearbox division was of equal importance and vital to Britain's military prowess.

The Merritt-Brown Gearbox

The tank transmissions became more than just an important part of David Brown's wartime contribution and were integral to the organisation's recognition as a leading defence equipment supplier, a role that continues even today. The company's involvement in tank gearboxes dates back to the 1930s and was largely the work of one of its research engineers, Dr Henry Merritt. Merritt's investigations into the theory and design of gearbox and steering systems for armoured fighting vehicles eventually led to the development of a regenerative steering mechanism incorporated into the tank transmission.

Henry Edward Merritt, better known to his

TOP LEFT:
Testing Dowty hydraulic pumps, which were used to raise and lower aircraft undercarriage gear. The pumps were manufactured alongside the gears in the Aero Block.

TOP RIGHT:
The Aero Block survived attempts by the Luftwaffe to pinpoint its location and still exists today. Traces of the camouflage paint it wore during the Second World War can still be found.

LEFT:
A convoy of tanks, including two Valentines and a Matilda, arriving at Park Works for gearbox trials. David Brown and Sons had been developing tank gearboxes since 1935. The first two vehicles carry the insignia of the 8th Armoured Division while the policeman escorting the convoy looks rather incongruous in the Daimler Dingo scout car.

colleagues as 'Ted', was undoubtedly a genius in his field and thought by some to have been one of the finest gear engineers in the world. During his lifetime, he presented several papers on transmissions to the Institution of Mechanical Engineers, and his books on gear technology became the standard works on the subject. His only failings were that he seems to have been a theorist rather than a practical engineer, and that his good sense of humour was often masked by his haughty demeanour, which did not always

ABOVE:
The Merritt-Brown tank gearbox. Designed by Dr Henry Merritt, it was a triple-differential transmission with a regenerative steering system.

endear him to all who worked with him.

Merritt was born in 1899 at Leyton in north London, strangely enough only just up the road from Albert Kersey's birthplace in West Ham. In 1915, when he was sixteen years old, he joined the aircraft and armaments manufacturer, Vickers Ltd, to serve as an engineering apprentice at the company's Erith Works on the marshy shores of the Thames estuary. He was eventually employed by Vickers as a tool draughtsman, before leaving in 1920 to take up a teaching post at Witham Technical College in Essex.

Between 1920 and 1924, Merritt worked as an assistant lecturer in engineering at Witham while working towards his masters degree, which he took in 1923; this was followed by his doctorate in engineering at London, two years later. In 1925, he joined David Brown to work as a research engineer under Arthur Sykes at Park Works. During his time at Huddersfield, he travelled abroad extensively, visiting Europe, the USA and the USSR in connection with his investigations into the development of transmissions for a wide range of transport and industrial machinery.

The traditional method of steering a tracked vehicle was by disconnecting the drive to one track either through a clutch and brake or a braked differential (as used on the first British tanks from 1915) mechanism. Both ways were inefficient and resulted in power loss. The concept behind the regenerative steering system was to incorporate the steering mechanism into the transmission, integral with the differential and gearbox. The drive to each track was controlled by the action of the gears, allowing smooth turns under power. The steering levers changed ratios in the transmission to steer the vehicle thus recirculating the power.

Preliminary experiments with geared steering systems, attributed to Major Walter G. Wilson, a pioneering engineer involved in the design of the First World War tanks dated back to 1916. Eventually Wilson had perfected a controlled double-differential system with two three-speed epicyclic gearboxes in place of the normal steering clutches. This transmission was fitted to the prototype Vickers A-6 '16-tonner' tank in 1928.

Wilson's transmission was the precursor to the regenerative steering system, but the design was limited by its over-complexity. While at Park Works, Merritt devised a more practical double-differential steering mechanism for tanks, which was patented by the David Brown company in 1935. His design was allegedly influenced by a French arrangement for the Char B tank that dated back to 1929.

Merritt's work soon came to the attention of the War Office and he was asked to join the Tank Design Branch of the Royal Arsenal at Woolwich as superintendent, leaving Park Works in May 1937. He was joined at Woolwich by Vernon Cleare from the Mechanisation Experimental

Establishment at Farnborough. Together they worked on a controlled differential transmission based on Merritt's double-differential steering mechanism mated to a seven-speed Maybach gearbox.

The transmission was evolved in conjunction with Maybach Gears Ltd, the British subsidiary of Maybach-Motorenbau of Friedrichshafen. This German industrial concern, formed in 1909 by Count Zeppelin and Dr. Maybach, was well known in the automotive field for its engines and gearboxes.

Maybach's British representative, Ernst Schneider, a respected engineer in his own right, assisted with the design, which became known as the Merritt-Maybach transmission. Schneider, a German national, returned to his homeland just before hostilities broke out. That a department of the British government would let a German subject work on a sensitive new design for a tank transmission is almost beyond belief and would later have serious repercussions.

The transmission was actually tested by speed ace, John Cobb, on his Railton Mobil Special car (which broke the land-speed record in 1939 with a speed of 369.74 mph) before being fitted to an experimental A-16E1 tank (top speed 30 mph), but neither the tank nor the gearbox went beyond prototype stage. While on the subject of land-speed records, it is interesting to note that Sir Henry Segrave, Captain George Eyston, Sir Malcolm and Donald Campbell all used David Brown gear units in their record-breaking cars.

As the preparations for war were stepped up in early 1939, the activities of the Tank Design Branch and the Mechanisation Experimental Establishment were amalgamated to form the Department of Tank Design. Henry Merritt was appointed assistant superintendent of this new combined section, taking over the responsibility for the existing gearboxes that were fitted to the latest British cruiser tanks, which incorporated features from Wilson's designs using two-speed epicyclic steering units.

The Covenantor tank, which was manufactured by the London, Midland & Scottish Railway Company at Crewe, had a Meadows-Wilson gearbox, while the Crusader, built under the auspices of Nuffield Mechanisations & Aero Ltd, had the largely similar Nuffield-Wilson gearbox. The transmissions were an improvement on earlier designs, but were still inefficient and dogged by reliability problems.

A contract for the manufacture of 1,400 Meadows-Wilson gearboxes, including the experimental and pilot-build models, for the Covenanter tank was awarded to David Brown & Sons. The company had been building tank gearboxes for several years as part of the work carried out by its heavy-engineering section.

Since 1938, the heavy-engineering department at Park Works had been in the capable hands of Fred Marsh. Born in 1905, Frederick Boyer Marsh served his apprenticeship with J. & H. McLaren of Leeds. During his time with this famous steam engine builder, he took his diploma in mechanical engineering at Leeds University. He worked for a couple of other leading Yorkshire engineering concerns before joining David Brown & Sons as a research engineer in 1936.

In the lead up to the war, Marsh was given the responsibility of establishing a new tank gearbox division at Park Works. One of his best draughtsmen, Herbert Ashfield, was entrusted with the gearbox designs. Ashfield, born in 1914, had joined the company as a student apprentice. After obtaining a diploma in engineering at Huddersfield College of Technology, he was seconded to the drawing office to join Arthur Sykes's team of design draughtsmen.

In October 1939, Merritt began working on a new and much improved triple-differential transmission, drawing on the expertise of a number of engineers within the Department of Tank Design, principally Lieutenant Colonel Ewen G. M'Ewen, Professor Hartree and Mr W. Steeds. While Merritt dealt with the theory, his assistant, Cleare, turned the calculations into a schematic drawing.

Plans were sufficiently advanced by December for the responsibility of preparing detailed drawings and constructing prototype transmissions to be handed over to David Brown. The company was the obvious choice; not only did it have a lot of experience with tank gearboxes, it was also Henry Merritt's former employer and held the patent rights to his original designs.

Once war broke out and the defence contracts

The David Brown Tractor Story

RIGHT:
The heavy machine shop for the tank gearbox division at Meltham Mills. The division moved from Park Works in late 1939 ready for production to begin at Meltham during 1940.

BOTTOM LEFT:
The internal components of a Merritt-Brown gearbox, which was manufactured at Meltham from July 1940. This is the production H4 transmission for a Churchill tank.

BOTTOM RIGHT:
Designated the A-22, the Churchill tank was the first armoured vehicle to be fitted with a Merritt-Brown gearbox. The regenerative steering system incorporated into the transmission allowed the driver to vary the radius of the turn in relation to the gear engaged.

poured in, Park Works was at full capacity with little or no room for expansion. Therefore, Fred Marsh decided to relocate the whole of the tank gearbox division, including the drawing office, to Meltham Mills. The move began at the end of 1939, taking over three long sheds that were still being used for making bobbins. The sheds, which were at the upper end of the mill site towards the top of Knowle Lane, were cleared out and tank gearbox production was phased in during 1940. The buildings were on two levels with the assembly on the upper level and the heavy machine shop on ground floor. Arthur Blackwell was appointed the chief works superintendent; the assembly foreman was Stanley Coats, and Edgar France was the machine-shop foreman.

An experimental shop, run by Jack Booth, was established in another old bobbin mill close to the tank gearbox works. It housed facilities for testing the tank transmissions and a machine-tool repair shop. Herbert Ashfield became the tank gearbox division's chief draughtsman and was put in charge of the drawing office. This remained separate to the tractor design department, which was still run by Albert Kersey. Ashfield worked closely with Merritt on the gearbox designs and suggested several changes and improvements.

The triple-differential transmission, which became known as the Merritt-Brown Type-301C was a logical development of the regenerative steering system. It allowed the driver to vary the radius of turn in relation to the gear engaged - the lower the gear, the tighter the turning circle. The transmission was also designed so that in neutral the tank would pivot on its own axis. The idea was to keep the radius of turn compatible with the travelling speed so that the vehicle remained stable at all times.

LEFT:
Churchill tanks under construction at the Vauxhall Motors factory in Luton. The power-plant for the tank, a 21 litre horizontally-opposed twelve cylinder unit, was designed by Alex Taub who had acted as consultant on the David Brown tractor engine.

The first prototype Merritt-Brown units were ready by early 1940 and one was fitted to the A-20, an experimental tank built by Harland & Wolff, for trials at Woolwich. Another prototype transmission was installed into a Covenanter tank, which was loaned to David Brown to act as a test bed for the new gearbox.

The Covenanter was delivered by road to Park Works, but as there was nowhere to keep it there, it was decided to send it to Meltham Mills. Fred Meadowcroft was given the job of driving the tank from Lockwood to Meltham. He recalls being summoned to the manager's office and being asked if he could drive a tank, to which he replied:

'Drive one? I've never even bloody well seen one!' Even so, he managed to fire up its Liberty aero-engine and master its tiller-steering system, and the journey was completed without incident. The tank was used for field trials out on Crosland Moor. Park Works' engineers under William Tuplin also built a rig to Henry Merritt's requirements that allowed the transmissions to be tested under load in the workshop.

The Merritt-Brown transmission went into production in July 1940, and its first application was in the new Churchill tank. The Churchill, which had been initiated by Merritt, was a development of the experimental A-20.

LEFT:
The Churchill tank went into production in the summer of 1940 but early models were dogged by reliability problems caused by engine and transmission failures. The gearbox problems were rectified by David Brown's chief draughtsman, Herbert Ashfield.

Designated the A-22, it was built by David Brown's old friends at Vauxhall Motors in Luton.

Vauxhall was awarded the contract after Alex Taub claimed in March 1940 that he could produce an engine larger and more powerful than anything the company had ever made before for the proposed tank from scratch in less than ninety days! The prototype power-plant, a 21 litre horizontally-opposed twelve-cylinder unit with side valves, was running by 11 June and had been designed and built on schedule with one day to spare. When tweaked and connected to the dynamometer, it put out over 350 hp.

The retreat to Dunkirk left the British Army seriously short of tanks, and the Churchill was rushed into production before many of its defects were ironed out, not least problems with its engine and transmission. Taub's Bedford Twin-Six, as the engine became known, proved unreliable, while the Merritt-Brown box caused one or two runaways under full power. Herbert Ashfield was given the job of troubleshooting the gearbox.

Merritt took Ashfield with him to a meeting with Winston Churchill to discuss the tank's shortcomings. It was the meeting where Churchill was famously to say of the tanks that carried his name: 'We'll make them anyway, and if they turn out to be no good, we'll put them on the cliffs of Dover and use them as pill-boxes'. Most of the defects were eventually rectified, but not before a number of the tanks had already been sent out to the Middle East.

Many of the problems with the early gearboxes were due to leakage from both the oil seals and the engine cooling system. Fluids mixed with brake dust formed a gunge that clogged the gear mechanism, jammed the steering and led to some spectacular transmission failures. Sometimes if the steering mechanism would not release when turning, inexperienced drivers, instead of stopping, tended to grab the opposite steering lever, bringing two drives into play at once and causing the gearbox to explode!

Herbert Ashfield tried out different materials for the oil seals and changed the double-helical gears used in the early gearboxes to stronger straight-toothed spur-gears. The early Churchill gearbox, a five-speed unit, was eventually replaced by an improved four-speed transmission known as the Merrit-Brown H4/1. The 'H' designation was given to horizontal versions of the Merritt-Brown box, while vertical models were prefixed 'Z'.

Once the initial teething problems with the Merritt-Brown box were sorted, it proved to be a reliable and revolutionary unit that became central to all the wartime British tank designs. Variants of the gearbox were fitted to a variety of armoured fighting vehicles, including the production Centaur, Cromwell, Comet and Challenger tanks, the Avenger self-propelled gun and the prototype Tortoise - a tank-destroyer and the heaviest fighting vehicle ever made in Britain. The Z5 (vertical five-speed) version of the transmission was very successful and went on to have postwar applications.

Henry Merritt had returned to David Brown at the end of 1940, having become disillusioned with the political infighting at the Department of Tank Design. Merritt was made the technical director at Meltham and given the responsibility for the design and development of both tank transmissions and farm tractors. By January 1941, the tank gearbox division was employing over 150 people. The following year, it became officially part of David Brown Tractors Ltd and Fred Marsh was appointed commercial manager for both undertakings.

It was impossible for David Brown to build all the gearboxes that were needed, particularly after Lord Beaverbrook joined the Ministry of Supply in 1941 and gave as much impetus to tank production as he had to the aircraft industry. Consequently, the Merritt-Brown units were eventually produced by a number of other manufacturers, including Leyland Motors, Nuffield Mechanisations, Armstrong Siddeley, Dennis Brothers and ENV Engineering. David Brown concentrated mainly on the Churchill and Cromwell transmissions, but remained the parent organisation, responsible for the gearbox's design and co-ordinating its production, while all the other concerns worked under its direction.

Henry Merritt and Herbert Ashfield made regular visits to the firms producing the Merritt-Brown gearbox to offer advice or solve an manufacturing difficulties. The two most prolific manufacturers involved were Nuffield Mechanisations and Leyland Motors. Nuffield was producing tanks at its Washwood Heath factory, adjacent to its Wolseley Motors plant in Birmingham. Merritt had a good working relationship with the Nuffield Organisation's vice-chairman and managing director, Sir Miles Thomas, who was also chairman of the government's cruiser tank production panel. Leyland's tank factory was in the town of Leyland in Lancashire, and it produced nearly 6,000 Merritt-Brown gearboxes between 1942 and 1945.

Nothing, however, overshadowed David Brown's contribution. On top of manufacturing new transmissions, the tank gearbox division at Meltham also reconditioned a very large number of units. As the work increased, overspill assembly expanded into the redundant worsted mill at Scarr Bottom.

The Ministry of Supply ensured that at least one tank was always made available to David Brown for test purposes. Usually delivered to Meltham by rail, they were unloaded in the marshalling yard at Scarr Bottom and the locals became accustomed to seeing them trundle through the village. The tanks were used as test beds for new gearbox components, and field trials were conducted out on the moors around Meltham.

For the first couple of years, the tanks were put through their paces by David Brown's own staff – they might have had no formal instruction but a least they knew how the gearbox operated. As the war progressed, the ministry insisted on providing its own personnel for the job. These army drivers could handle a tank, but they didn't know the moors as well as the Meltham men, and at least one of the armoured vehicles had to be ignominiously recovered from a ditch – much to the amusement of the David Brown employees!

Because of the sensitive nature of the work that was carried out at Meltham Mills, uniformed factory policemen were always stationed at the gates while the home guard patrolled any deserted buildings at night. Even so, there was still the continual worry about saboteurs and spies. For this reason, whichever tank was allocated to Meltham Mills was always secreted away in its own storage shed on the site when not in use.

Sometimes when there was more than one tank on site, the spare was often hidden in outbuildings at Durker Roods. This caused problems of its own one evening when officials from the Inspectorate of Fighting Vehicles came to look at the spare tank and found the outbuildings empty. It seems Mr David Brown had decided to take the tank out on the moor and was currently doing a circuit of Holme Moss all on his own!

BELOW:
The tank gearbox assembly floor at Meltham. The division which was housed in former bobbin mills produced over 10,000 transmissions and reconditioned another 6,000 during the Second World War.

RIGHT:
A Comet tank fitted with a Merritt-Brown Z5 gearbox leaving the Leyland Motors factory in Lancashire. As the war progressed, other manufacturers produced the tank gearbox under David Brown's supervision, including Leyland who built nearly 6,000 Merritt-Brown transmissions between 1942 and 1945.

RIGHT:
Vertical models of the Merritt-Brown gearbox were prefixed 'Z'. This five-speed Z5 version was fitted to the postwar Centurion tank.

In 1943, Herbert Ashfield was invited to inspect a gearbox that had been taken out of a German Tiger tank, which had been captured in the Western Desert that April after being knocked out, quite fittingly, by a Churchill. He was amazed to find that the transmission was very similar to the Merritt-Brown box and almost identical to the earlier Merritt-Maybach double-differential system. It seems that Herr Schneider had indeed scuttled back to Berlin with the plans for Britain's new regenerative steering mechanism under his arm!

By the time the war came to a close, Meltham had built some 10,000 tank gearboxes and reconditioned or adapted another 6,000. It was not the end of tank gearbox production because the Merritt-Brown Z5 box was incorporated into the post-war Centurion and Conqueror tanks.

Tank transmission production continued at Meltham for several years after the war until the limitations on space due to an expansion of the tractor operations prompted a move to a factory at Heckmondwike, near Batley. The Heckmondwike factory, known as Orchard Mills, was a former textile mill that had been requisitioned as a store by the Ministry of Food during the war. In 1950, David Brown Tractors Ltd took a lease on the premises and used it as a temporary parts store before establishing a new heavy gearbox division for tank transmissions there two years later.

The division at Heckmondwike was set up under the auspices of the Ministry of Defence and was managed by Harry Pilkington. The prototype tank gearboxes were still made in the experimental shop at Meltham. David Brown Tractors Ltd remained the parent company for the division and the profit from this undertaking contributed considerably to the tractor research and development budget.

During the mid-1960s, tank gearbox production reverted to Park Works. In later years, Merritt's steering system was grafted on to an epicyclic speed-change mechanism. This new gearbox, designated the Merritt-Wilson, became the basis of the transmissions for the Chieftain and Challenger main battle tanks, as well as the Scorpion light tank, with David Brown remaining the principal contractors. Incidentally, during the late 1970s Herbert Ashfield was brought out of retirement by David Brown Gear Industries Ltd and was asked to solve the same tank transmission

ABOVE:
David Brown still builds tank gearboxes today at Park Works and supplied the six-speed TN54 automatic transmission for the latest Challenger 2 that saw service in the recent Iraq conflict. Another firm with agricultural connections, Perkins Engines of Peterborough, supplied the tank's V-12 diesel power-plant.

problems that he had been faced with forty years earlier!

Even by the early 1990s, the order book for David Brown's Military Products division was still almost full. Park Works had contracts for the production of the six-speed automatic TN54 gearbox for the Challenger 2 tank for both the British Army and the Shah of Oman, as well as overhaul work on the European-based US Army MLRS and Bradley gearboxes. In 1995, the Ministry of Defence's follow-on purchase of 259 more Challenger 2 tanks further accelerated production of the TN54 transmission. The transmission is still made at Park Works today, having lately proven itself in the battle-hardened conditions of the recent Iraq conflict.

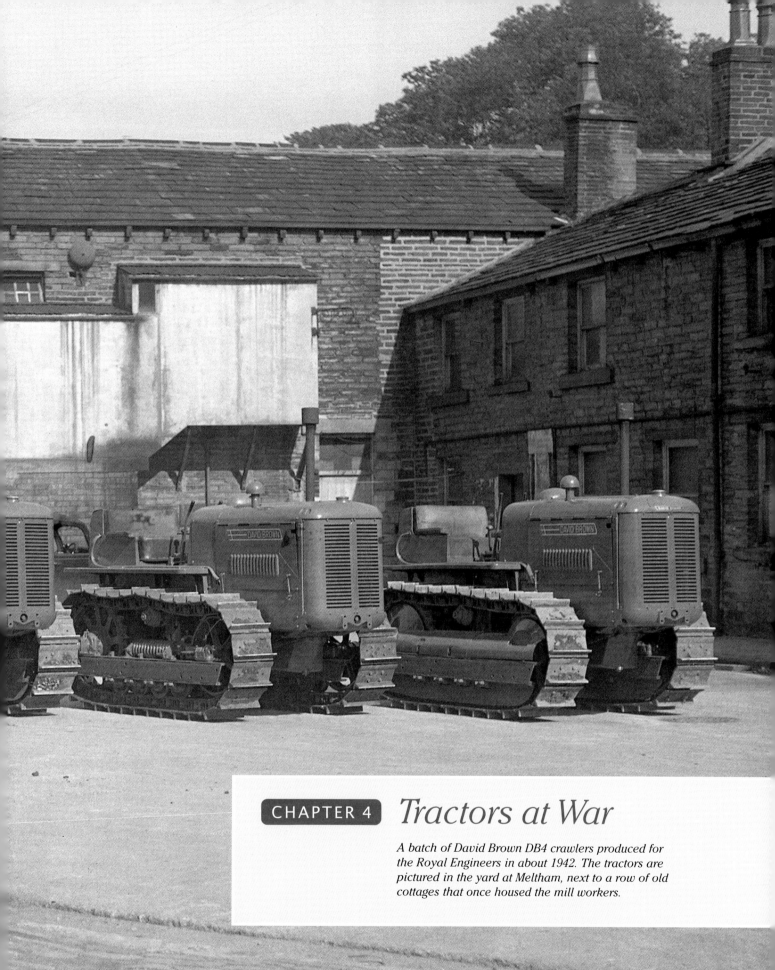

CHAPTER 4 — *Tractors at War*

A batch of David Brown DB4 crawlers produced for the Royal Engineers in about 1942. The tractors are pictured in the yard at Meltham, next to a row of old cottages that once housed the mill workers.

With everything else that was going on at Meltham during the war, it would be easy to forget that the company still found time to build tractors. But build them it did, and in no small numbers. By 1945, somewhere in the region of 7,500 David Brown tractors of various different sorts had rolled off the production line. It might have only been a small fraction of what the mighty Ford organisation (which had been awarded the main contract for the supply of agricultural tractors) had produced at Dagenham in the same period of time (just under 138,000 Fordsons), but it remains a commendable effort considering that most of Meltham's resources had been concentrated in other directions.

As would be expected, tractor production at Meltham Mills had begun very slowly - the move to the new site in the summer of 1939 being overshadowed by world events. The logistics of setting up new assembly lines in buildings more suited to spinning and weaving were complicated by most of the company's facilities being tied up with defence contracts. Many of the machine tools and jigs for the tractor division were made in the tool-room at Park Works under the supervision of Harry Davies.

Albert Kersey moved to Meltham to take up the position of chief engineer of the tractor division. Bill McCaw also went to the design office at Meltham for a short time but left at the end of 1939 to join Westland Aircraft at Yeovil in Somerset. The new chief draughtsman for tractors, James Stirling, came from Vauxhall Motors. Most of the engineering staff were either ex-Karrier personnel or were recruited from Morris or Vauxhall.

Several of the other positions within the tractor division were carried over from the Ferguson-Brown days. Walter Hill was still general manager and Harry Pilkington was works manager. Arthur Blackwell, as chief works superintendent, was also in charge of tractor production, with Fred Armitage and Herbert Walker as his line foremen. The line inspector was Emil Moes, a Belgian who had joined Park Works as a fitter in 1916.

RIGHT:
The No.1 machine shop for the tractor works at Meltham Mills. Here, the rough castings, including the single-piece mainframe and the rear-axle housings, were machined ready for final assembly.

Captain Arlborough-Smith took charge of tractor sales and ran the personnel department, but left the company towards the end of the war to take up an appointment with an implement manufacturer. The tractor division's two Austin 10 vans, four Morris 8 cars and Bedford OY truck were looked after by the service department run by Bill Wallis, assisted for a time by Len Craven. A temporary stores with a few bins for parts was established on the second floor of an old garage attached to Wood Cottage, opposite the works main gate on Knowle Lane. The garage later became the works fire station.

ABOVE:
Part of David Brown Tractors' wartime transport fleet, two Austin 10 service vans and a Bedford OY 3 tonner, outside the Aero Block on Meltham Mills Road. Note the camouflage on the building.

LEFT:
An ARP warden with a David Brown tractor and a fire-pump tender outside the works fire station housed in an old garage that was part of the remains of the Clock Mill on the corner of Knowle Lane and Acre Lane.

RIGHT:
One of the first production tractors leaving Meltham plant from the gate on to Knowle Lane next to the main office block. The articulated truck is a Scammell Mechanical Horse that was operated by the London Midland Scottish Railway.

BELOW:
A differential unit is fitted on the assembly conveyor in the tractor works at Meltham Mills. Most of the main components were manufactured on site.

Tractor Developments

Records show that probably no more than forty-seven tractors were built during 1939. Production expanded dramatically the following year, but still remained in line with the Ministry of Supply controls. These restricted volumes of agricultural tractors to around 1,000 per year, and production remained at this level for the duration of the war. It would be 1941 before the plant was finally fully equipped for tractor manufacture with the introduction of a moving conveyor for the main assembly line and the inception of an Asquith multi-purpose machine with four heads for machining the rear-axle housing in one operation.

Most of the main tractor components were manufactured in house on the Meltham site, which housed not just the assembly and feeder lines but also all the machine shops and heat-treatment works related to tractor production. Ancillary parts were sourced from outside manufacturers, such as the Automotive Products Company, which supplied the Borg & Beck clutch and the Thompson self-adjusting tie-rod. The radiator came from H. O. Serck Ltd of Manchester and the brake components from Girling Ltd of Birmingham. The carburettor was a Solex vertical type 30-FV and the fuel pump was supplied by AC-Delco.

The actual price of the Davi

LEFT:
The finishing end of the tractor assembly conveyor at Meltham. Production volumes were restricted to around 1,000 agricultural tractors per year during the war.

Brown tractor in basic form on steel wheels was set at £189. It was also available with 9.00 x 24 (rear) and 4.50 x 19 (front) 'General Purpose' pneumatic tyres for £210. For an extra £15, the tractor could be specified with 9.00 x 28 (rear) and 5.00 x 19 (front) pneumatics. The tyres were supplied by Firestone. While on the subject of wheels, it is worth noting that David Brown Tractors Ltd pioneered the use of dished wheel centres to alter track width - just one of many 'firsts' patented over the years by the company.

The new tractor was offered with a choice of

LEFT:
The David Brown tractor in basic form on steel wheels was priced at £189 in 1940. The photograph was taken from the works playing fields.

RIGHT:
A restored 1940 David Brown tractor. This machine would have originally cost £210 on pneumatic tyres. David Brown was the first company to pioneer dished wheel centres to alter track width.

electrical equipment from simple magneto ignition to a full 6 volt starting system using a CAV battery and fuse box with a Lucas starter motor and dynamo. Some of the very early tractors had Lucas magnetos, but most were fitted with Scintilla Vertex NV4 units. Lighting equipment was also available as an optional extra. Models with electric starting and lighting were normally equipped with Lucas/CAV distributor and coil ignition.

A fixed 'U'-shape drawbar, known as th 'geometric drawbar', was supplied with the tracto for trailed implements. A combined belt-pulley an power take-off unit was also available and bolted t the rear-axle housing in place of the differential cove

RIGHT:
The David Brown was available with oversize tyres, described as 'Super Traction Pneumatics', for an extra £15. The tractor is a 1941 model.

The first price list and sales brochure for the David Brown tractor still listed the hydraulic unit as being available for an extra £30. Initially, customers who had bought the tractor and were waiting for the hydraulic lift had been told that the unit would be available by early 1940, but this proved not to be the case. In reality, it had not been released for production because it had been discovered that the original unit infringed Ferguson's patents and there had been no time to develop a reliable alternative.

To satisfy the clamour, particularly from Northern Ireland, for implements to suit the tractor, David Brown looked into the possibilities of producing a trailing plough as an interim measure until the power lift was ready. The Oliver plough that had been used for the initial field trials on Crosland Moor was delivered to Meltham for the draughtsmen to look at while Fred Meadowcroft reported on its performance to Albert Kersey.

Kersey based his design on Meadowcroft's evaluation, possibly incorporating a number of Oliver, Fisher-Humphries and Ransomes Motrac features into the plough, which was fitted with Ferguson bodies. The David Brown trailing ploughs were designated PW1 and a batch of 500 was built in the basement of the tractor assembly

ABOVE:
A David Brown tractor with an experimental drawbar on trial in 1940. Problems with finalising the design of the hydraulic lift meant that the early production tractors could only handle trailed equipment.

LEFT:
A batch of 500 trailing ploughs was built in 1940 to meet the demand, particularly from Northern Ireland, for implements to suit the David Brown tractor. The ploughs cost £35 each with a choice of general-purpose or semi-digger mouldboards.

RIGHT:
The prototype U36 power lift. Developed during 1940, it was a self-contained unit that bolted on to the tractor's rear-axle housing and incorporated a power take-off drive.

BELOW:
The David Brown tractor with the U36 power lift and the PU1 two-furrow plough with general-purpose bodies. Early trials uncovered a tendency for the tractor's parallel linkage to allow the plough to crab rather than follow in a straight line.

shop. They cost £35 each with the option of general-purpose or semi-digger mouldboards.

By July 1940, Walter Hill felt compelled to issue the following statement: 'We can only apologise for the temporary hardships occasioned to those farmers who purchased David Brown tractors expecting that the power lift unit would follow in time for inter-cultivation this spring, but under the present circumstances they will realise that we must work to priority instructions and to schedules laid down by the Government Departments.'

Government contracts did take precedence, but Kersey's engineers were still exploring the possibilities of circumventing Ferguson's patents and had experimented with both hydraulic and mechanical power lift units. By March 1940, the design of a hydraulic unit, provisionally designated U36, was beginning to take shape. Like the earlier lift, it was a self-contained unit that bolted on to the rear-axle housing and included a power take-off drive. It was, however, a more compact unit with a single ram-cylinder and did not incorporate any sort of hydraulic depth or draft control system so as to avoid Ferguson patents.

The U36 unit still had the same type of four-cylinder reciprocating pump as fitted to the Type A tractor because there were still a considerable number remaining in stock from the Ferguson-Brown days. The Ferguson patents were circumvented by moving the rotary spool-type control valve from the suction to the delivery side of the pump. Unfortunately, this was not ideal because the oil heated up under pressure, but it was the best solution open to the engineers at the time.

Another Ferguson patent covered the converging three-point linkage, so Kersey's team designed the lower link-arm mountings to be the same distance apart as the attachment points on the plough's cross-shaft. This meant that the lower links remained parallel and there was no patent infringement.

When Charles Hull joined Kersey's team, one of the first jobs he was given was checking the component-detail drawings for the U36 unit. In July 1940, Hull was asked to prepare an assembly drawing in readiness for the power lift being released for production. The first completed unit was ready by the autumn and was fitted to a tractor for trials, which were carried out by Bill Harrison in a field at Wilshaw, about a mile from the Meltham plant. Harrison reported back that the plough was uncontrollable, and Charles Hull (who believes he was given the job because he rode to work on a motorcycle and had the only pair of long rubber boots suitable for traipsing across ploughed fields) was sent out to investigate.

The problem was that the parallel link-arms could not control the side-draft of the plough in the same way that Ferguson's converging linkage could. The implement tended to crab rather than follow in a straight line behind the tractor. Charles Hull came up with the idea of fitting the plough with a rocking cross-shaft or zed-bar that was linked to the inside of the nearside lower link-arm. If the link-arms swung sideways, a compensating link, known as 'the diagonal link', pushed on the cross-shaft, angled the plough and steered it back into its correct position. The length of the diagonal link could also be altered to adjust the front furrow width. It was a simple but ingenious solution that paved the way for the U36 unit to go into production.

BELOW:
Trials of a David Brown tractor and plough fitted with the diagonal linkage, which can just be seen fitted between the implement's cross-shaft and the left-hand lower link-arm. Designed by Charles Hull, it kept the plough in the correct position behind the tractor.

The new hydraulic unit was launched in early 1941 and cost £49 10s. When it was fitted to the tractor, some parts of the standard drawbar became surplus. As a concession to those farmers who had been waiting some time for the power lift, a payment of £1 10s was paid for the redundant parts providing they were returned to the works in good condition. This concession continued until February 1946, after which time all hydraulic units were factory fitted.

ABOVE:
An early David Brown PU2 two-furrow plough with semi-digger bodies. The bracket for the diagonal link can just be seen beneath the cross-shaft.

RIGHT:
The CR3 nine-tine rowcrop cultivator. The implements in the David Brown range of mounted equipment, launched in 1940, were mainly constructed from surplus Ferguson-Brown parts.

The power lift's availability enabled David Brown to offer a range of 'unit principle' mounted implements. These had been designed for some time, and relied heavily on Ferguson-Brown components because the company had quite a stock of surplus parts in store. The plough range consisted of a 10in. two-furrow general-purpose model, a 12in. two-furrow semi-digger and a 16in. single-furrow deep-digger plough. A spring-tine 'dual-purpose' cultivator, a rigid-tine 'rowcrop' cultivator and a three-row ridger completed the implement line-up. The spring-tine was based on the Ferguson cultivator, but the arrangement of the spring mechanism was altered to avoid patent infringement and allow the obsolete parts to be used up.

As mentioned previously, because of Ferguson's patents, the power lift did not incorporate any form of hydraulic draft control. To counteract this, Fred Meadowcroft suggested the simple solution of providing all the implements with depth wheels. Conversion sets were also offered to allow the David Brown trailed plough, as well as the earlier Ferguson-Brown implements, to be used with the U36 power unit.

Following his work with the diagonal link, Charles Hull was promoted to the position of section leader for implements in September 1940. His first responsibility was to rectify defects in the lift mechanism for the trailing plough, which was affecting up to 75 percent of the models already sold. When the plough was tripped, a rack meshed with a pinion on the wheel and was raised into a self-locking position. Hull, trained in precision engineering, was amazed at how loose the tolerances were on the black-iron components.

'I was brought up to believe that a thou was as big

as a teacake,' he recalls. 'However, I soon realised that farmers left their implements out in fields over the winter, and if they had been manufactured with close tolerances, they would rust solid. It was what we called the sloppy-fit lubrication system.'

The faults with the lift mechanism were rectified by altering the angle at which the banana-shaped rack connected with the pinion. While testing the plough out on Crosland Moor, Fred Meadowcroft found the implement difficult to hook up to the tractor because much of the weight was on the drawbar. To counteract this, he developed a cone-shaped hitch, which made attaching the plough both quick and easy. Incidentally, in 1947 David Brown sold the trailing plough design to Bristol Tractors Ltd, then owned by the Austin car distributors, H. A. Sanders, of Earby, near Colne in Lancashire. It was marketed as the Bristol-Brown plough and cost £63 10s.

Henry Merritt arrived at Meltham at the end of 1940 and Bill McCaw was persuaded to return as chief draughtsman in January 1941 to develop a towing tractor for the Air Ministry. Two months later, Kersey called Hull into his office and informed him that he was being sent by train to Scotland to look into problems with the ridger. Hull, armed with his Stormguard motorcycle coat, was met at Aberdeen by representatives of the local dealers, Barclay, Ross & Hutchison, and taken to see the implement at work.

The ridger had what was known as a manual depth control unit, basically a depth wheel, fitted with a pneumatic tyre, that was mounted forward of the centre ridging-body. Problems were arising with farmers trying to use the implement to split

ABOVE:

The prototype David Brown ridger. All the implements were provided with a manual depth control unit, in this case a depth wheel fitted with a pneumatic tyre, to compensate for the tractor's lack of draft control.

LEFT:

The David Brown ridger at work. Early models gave problems in Scotland when they were used to split back the ridges after planting.

ABOVE:
Following investigations by Charles Hull in early 1940, the three-row ridger was fitted with a new type of depth control unit incorporating two dished wheels. The wheels could be reversed in the form of a diabolo to grip the ridges when splitting back.

RIGHT:
David Brown offered this tractor canopy from April 1941, priced at £7 10s. It was a crude structure, but it did give the driver some protection from the harsh Yorkshire weather.

back the ridges after planting seed potatoes and finding it impossible to keep the depth wheel on top of the row. Hull returned to Meltham and consulted with Merritt and McCaw who came up with the solution of a manual depth control fitted with two dished wheels in the form of a diabolo to grip the ridge.

The David Brown tractor was ideal for rowcrop work because of its offset driving position; rather uniquely, the operator sat with both feet on the right-hand side of the transmission housing. There were two pedals - the left operated the clutch, although a hand-clutch lever was also provided. The rear wheels had independent brakes, which could be operated together by the right pedal, linked to both drums via a cross-shaft, or worked separately by levers on either side of the bench seat for turning on headlands.

David Brown was one of the first manufacturers to offer some form of protection for the driver when it launched its tractor canopy in April 1941. Priced at £7 10s, it was a basic structure consisting of a canvas cover on a steel framework. Its introduction was no doubt prompted by the harsh winters experienced on the Yorkshire moors, although the press release accompanying its launch suggested it was needed because of the heady speeds the tractor could achieve on the road. It read, 'David Brown tractors ... have an alternative governor setting, which gives a road speed of 20

miles per hour. At this speed, particularly in inclement weather, the canopy will be particularly valuable.'

A top speed of 20 mph may seem nothing out of the ordinary by today's standards, but it was unusually fast for the 1940s and gave the David Brown tractor a useful edge over its competitors when it came to road work. The Ferguson-Brown had been flat-out at 5 mph, while the Land Utility Fordson could only manage 8 mph. Even the company's Austin 10 service vans would only go 10 mph faster!

The David Brown tractor's top speed was attained in the second of the two governor settings. The two settings were sometimes limiting and some farmers would have preferred an all-speed governor. Criticism was also levelled at the vaporiser because it was thought that this could have been more efficient in allowing a quicker changeover from petrol to vaporising oil.

Vaporising oil, a derivative of paraffin or kerosene and better known as TVO (tractor vaporising oil), was a heavy distillate that would only convert into an inflammable vapour when hot, hence the need to warm the engine up on petrol. The more efficient the vaporiser, the quicker the changeover from petrol to the less volatile and cheaper TVO. This became increasingly important with wartime shortages, which led to petrol rationing. Changing over too soon from petrol resulted in misfiring with unburned TVO remaining in the cylinder leading to possible dilution of the lubricating oil.

To counteract this, Albert Kersey was charged with developing a combined inlet and exhaust manifold that would utilise the heat from the exhaust gases to allow a quicker warm-up and minimise petrol consumption. Because tractor development was not officially sanctioned due to wartime restrictions, much of the work was carried out in secret in a room above the tank gearbox experimental shop. Access to the room was up a rickety wooden staircase, which was not very practical, and so the decision was made to set up a tractor experimental department with a dedicated workshop in an empty building behind the tractor assembly works. A spare Heenan & Froude water-brake dynamometer was procured from the engine-test bay and installed in the building together with cooling and exhaust systems and fuel storage facilities. A rolling road was made from a pair of old steel wheels sunk into a pit.

Because tractor production was viewed as non-essential work, making it virtually impossible

BELOW:
A trio of David Brown tractors working in Worcestershire in 1940. The David Brown model was praised for its economy and reliability, and farmers found it far less temperamental than its competitor, the Fordson.

RIGHT:
A 1940 sales leaflet for the David Brown tractor.

to obtain permits for test equipment, Kersey gave Charles Hull the job of designing a fuel-flow meter for the dynamometer, which he devised from a two-way tap and a carburettor float. Basic workshop and machine tools, including two centre-lathes, a milling machine and welding and brazing equipment, were foraged from around the works. A makeshift strobe to test ignition timing was even made from pieces of scrap and a child's lantern. A brazing hearth and a small laboratory to analyse TVO were set up and the new experimental department was in business.

There was also an experimental tool-room at Meltham, which had been transferred from Park Works in late 1939 and occupied part of the tractor assembly line before moving into the tank gearbox experimental shop. It was mainly concerned with developing machine tools and the assembly of a few experimental tank transmissions, but it did build a number of parts for the tractor department.

The few criticisms levelled at the David Brown tractor paled into insignificance when compared to its obvious advantages. It was far less temperamental than the ubiquitous Fordson and was not prone to oiling its plugs so there was no need to keep a spare set warming on the manifold. It was also lighter to handle and easier to start. Comparison is probably unfair because Meltham's machine was a product of the 1940s and had the latest type of overhead-valve engine while wartime exigencies had forced Dagenham to carry on with an outdated design. Ford had been planning to replace its side-valve engine which had been in production since 1917 but all development work had to be shelved due to the outbreak of war.

Farmers praised the David Brown's dependable engine and marvelled at its economy. The tractor's performance was first highlighted after it was submitted to the Royal Agricultural Society of England's Tractor Testing Scheme with trials carried out over 20 and 21 December 1939. During an eight-hour ploughing test with a three-furrow Ransomes RSLM plough, it averaged 0.84 acres per hour returning a fuel consumption figure (on low-grade wartime 'Pool' fuel) of 1.61 gallons per hour. This equated to a very commendable result of 1.92 gallons per acre. The tractor was in second gear travelling at 3.2 mph. Its maximum drawbar-pull was measured at 3,200 lb.

Personalities and Problems

David Brown Tractors Ltd underwent a number of personnel changes during the wartime period. Walter Hill resigned in October 1940 (eventually retiring to Jersey after the war) and was succeeded as general manager by H. P. Scott, who had joined the company four months earlier. Scott, whose brother, Reg, ran Meltham's purchasing office, came to David Brown from Morris Motors where latterly he had worked as a production controller. Earlier in his sixteen-year

LEFT:
Mr. David Brown at the wheel of one of his company's aircraft towing tractors outside Meltham Pleasure Grounds. Sitting beside him is Henry Merritt, who rejoined David Brown as Meltham's technical director in late 1940.

career at Morris, he had been involved in the development of cross-country vehicles, a proposed farm tractor and, ironically, a cancelled venture with Harry Ferguson that had taken place prior to the Ferguson-Brown agreement.

While the parent concern, David Brown & Sons, remained a public limited company, David Brown Tractors Ltd was almost entirely owned by Mr David Brown, with the exception of £30,000 worth of preference shares used to finance its set-up. Father and son headed the board of the tractor company with Frank Brown as chairman and David as managing director. They were joined by Arthur Maxwell Ramsden, a partner in a Huddersfield firm of solicitors, Ramsden, Sykes and Ramsden. Max Ramsden, a well-known local figure, was both the Brown family's solicitor and a personal friend and confidant of David's. Bert Jenkins, who had recently joined the company as works manager, took a fourth seat on the board in 1941.

Henry Merritt also joined the board after he returned to David Brown as Meltham's technical director. Ostensibly, he was employed to oversee the development and production of tank gearboxes, but during 1941 he also took over the complete responsibility for tractor design. At first, the grounds for this seem unclear because Merritt was more of a theoretician than a practical engineer and had little knowledge of farm tractors. However, the appointment was precipitated by the death of Frank Brown from a stroke on 13 March of that year, which left his son with his hands full and, concerned by a seeming lack of progress in the design department, needing to leave his fledgling tractor division in experienced hands.

At the time, Mr David Brown was facing severe problems. He had taken the step of setting up an administrative board of four senior executives, plus himself as chairman, operating from offices at Meltham, to oversee the running of the David

BELOW:
Meltham Mills in 1942, possibly after that year's severe floods. The photograph gives a good idea of the differing levels of the site. The tractor works can be seen on the upper level with the machine shops to the left and the assembly shop to the right. Many of the Air Ministry tractors in the yard had either been flood damaged or brought in for reconditioning.

Brown group of companies during this crucial period. However, his father's death prompted an unexpected attempt at a boardroom coup by the four executives, who felt they were not being paid enough and were supported by David's uncle, Ernest Brown.

David felt isolated because his closest confidant, Max Ramsden, seconded to the army and promoted to the rank of brigadier, had been called away on urgent military matters. The rebellion forced the intervention of the Ministry of Supply, which suggested the appointment of an MP, S. S. Hammersley, as a caretaker chairman. It soon became evident that Hammersley was not up to the job and he was quickly replaced by Lord Brabazon of Tara, who had succeeded Lord Beaverbrook as Minister of Aircraft Production.

Brabazon had been a pioneer racing driver and saw Mr David Brown as a kindred spirit. He proved to be a staunch ally and loyal supporter of the managing director, and the attempted coup, which was ill timed, collapsed. On top of the boardroom struggles, Mr David Brown was also facing marriage problems and was plagued by severe migraines. It was a difficult time for him and he had to accept leaving Merritt to his own devices in charge of tractor developments at Meltham, even though his relationship with the engineer was often strained and never harmonious.

Another notable appointment at this time was Vincent Gallagher. A Yorkshireman, born at Settle in 1915 and educated at Giggleswick School, Gallagher had joined Park Works as a foreign correspondent in 1934. Moving to the purchasing department, he was involved in sourcing components for the Ferguson-Brown venture and was promoted to the position of chief buyer for David Brown Tractors Ltd at Meltham in 1941.

Henry Merritt's appointment caused some resentment within the tractor division, particularly after he demoted the respected Albert Kersey to the position of chief experimental engineer in October 1941. Kersey had been unfairly blamed for the problems with the parallel linkage and wrongly criticised for escalating costs and slow progress on tractor developments. In reality, the problems were due to a shortage of resources and a lack of testing because the

company did not have an organised field test and experimental department.

Kersey moved out of the design office into a small office in the new experimental workshop, taking Charles Hull with him as his assistant. Together they formed the Tractor and Implement Experimental Department, working on improvements to the existing tractor while constructing and testing prototype machines. They took over the responsibility for field testing both tractors and implements and brought Fred Meadowcroft into the department. A. E. Smith, previously an engineer with Blackstones of Stamford, was recruited as an implement designer.

Mr David Brown, who had given Merritt a brief to design a new tractor as a separate project to the work already being carried out on improvements to the original tractor design, was a little concerned about his technical director's lack of experience in automotive engineering. For this reason, he invited an esteemed colleague from the motoring world, Georges Roesch, to act as chief engineer to David Brown Tractors Ltd.

Roesch was one of the most revered engineers of the time with considerable experience of both motor car design and racing car developments. Swiss by birth, he had worked variously for Automobiles Gregoire, Delaunay-Belleville, Renault and Daimler before joining Clement Talbot, later Sunbeam Talbot, where he was chief engineer for twenty-three years from 1916 to 1939. He joined David Brown in May 1942.

Merritt had not been consulted about Roesch's appointment and was clearly unhappy about the situation, feeling it undermined his responsibility. Conversely, the Swiss engineer was not pleased to find that he and Merritt were both doing the same job.

Roesch worked on tractor engines, taking over the development of the new manifold. During the first few months of his time at Meltham, he made frequent visits to the experimental department, asking innumerable questions about the work in hand. He and Merritt tended to avoid each other and worked separately on their own projects. Eventually, the clash of personalities became too much to bear and Roesch left after less than a year to join the Ministry of Supply as chief mechanical engineer to the National Gas Turbine Establishment.

Steve Moorhouse, one of Herbert Ashfield's draughtsmen from the tank gearbox division who had joined David Brown after working in the engine drawing office at Bristol Aircraft, had assisted Roesch on the engine designs. A local man, Moorhouse had started his career with Broadbent's foundry before becoming a draughtsman at Woolwich Arsenal, where he prepared drawings of gas masks and pyrotechnics.

Rather craftily, the tractor division kept the best of its young engineers and draughtsmen on the payroll of the tank gearbox division, which was a reserved occupation, to avoid losing them in the call-up to the armed services. This led to confusion as some of the top draughtsmen worked on both tractors and tank gearboxes, under both chief draughtsmen, Bill McCaw and Herbert Ashfield, even though the two drawing offices remained separate at this time.

Georges Roesch was not the only engineer of some renown to visit Meltham during the wartime period because a number of consultants were brought in from time to time, some at Merritt's invitation and others at Mr David Brown's request, to advise mainly on engine matters. One such visitor was Edgar Sidney, the works manager of Bryce Ltd, renowned fuel-injection equipment specialists from Hackbridge in Surrey.

Sidney appeared at Meltham during the early months of the war and consulted with Albert Kersey on the development of a diesel version of the tractor engine. A few specialist parts, such as a heavy-duty cast-iron crankshaft and a heavy flywheel, were fabricated and a prototype power unit was constructed using a standard petrol/TVO block with Bryce fuel-injection equipment. The experimental engine was started up but was very smoky and Sidney found it would only run properly if he placed his fingers in the inlet ports! The project was eventually shelved when other developments took precedence. (After the war, Sidney became the technical director and works manager of another famous Surrey company, AC Cars Ltd of Thames Ditton.)

After Roesch's departure, Professor Wisner, who had been a research engineer with Citroen in

ABOVE:
The engineering staff from both the tractor and tank gearbox drawing offices in 1943. Left to right are: unknown; Dr Henry Merritt; Mr. Taylor; H. Deacon; unknown; Charles Hull; H. Samek; Stanley Mann; Bessie Nuttall; George Jemmison; Steve Moorhouse; unknown; Herbert Ashfield; Alan Whiteley; Arthur Turner; unknown; M. Green; Douglas McCann; D. Broadbent; unknown; Jean Marshall; Annie Bradshaw; Albert Kersey; unknown; Kenneth Dearnley; A. Knight; Mabel Kenyon; Irene Ellis.

France before the war, was engaged to act as a consultant to Henry Merritt. A few years earlier, the Frenchman had been involved in the development of a 40 bhp diesel engine for Citroen in collaboration with Harry Ricardo. Launched in 1937, it was designed for light-vehicle use and was one of the smallest automotive diesel engines on the market with a swept volume of only 1,766 cc.

Through the war, Wisner also acted for Simms Motor Units of London and Norris, Henty & Gardner, diesel engine manufacturers of Manchester on a consultancy basis and spent just two days a month at Meltham. Whether he advised on diesel matters or whether he briefed Merritt on other areas of engine design is not known. For a time his visits stopped, and it was rumoured that he had been secretly flown back into France to work for the clandestine Special Operations Executive in preparation for the D-Day landings.

Albert Kersey, who felt snubbed by Merritt, had become disenchanted with the tractor division and transferred to Park Works in July 1943 to work on gearbox and transmission designs, leaving Charles Hull to take over as chief experimental engineer at Meltham. Hull was a very capable engineer who was highly respected by his peers. Hailing from Leyland in Lancashire, he had naturally served his apprenticeship at Leyland Motors, joining the commercial vehicle manufacturer at the age of fourteen in 1931. During his time at Leyland, he worked in the research department under the great Stanley Markland, testing all types of engines including direct-injection and turbocharged diesel models and even a single-cylinder with poppet valves. This was followed by an eighteen-month spell with the Air Ministry at Farnborough, where he had hoped to work on aero-engines but was disappointed to find himself involved with compressed-air or cordite-propelled aircraft catapults.

In early 1940, David Brown Tractors Ltd advertised for a design draughtsman with diesel engine experience. Hull applied for the post, and after being interviewed by James Stirling and Captain Arlborough-Smith, he was offered the job with a salary of £4 10s per week. He joined the company in March and was immediately put to work on the design proposals for a single-cylinder stationary engine. After this was dropped, he moved over to the U36 project.

Charles Hull was assisted in the experimental department by Stanley Mann, who had previously worked for Norris, Henty & Gardner. Mann had joined David Brown as a draughtsman in the tractor drawing office in mid 1940, probably in response to the same advert that had attracted Hull. An apprentice from Park Works, Maurice Jones, was also seconded to the experimental department in 1943 to assist on the test bed.

Merritt's sojourn as technical director of David Brown Tractors Ltd was marked by his establishment of a series of important and strict engineering practices. Procedures that he instigated were handed down to his successors and ensured that the high quality of work in the field of tractor design continued at Meltham for decades to come. Each job in the engineering department was controlled by a project sheet, which defined the objective and purpose of the work. The project was allocated a number to which all costs were charged. The construction and testing of prototypes or components required written reports, which had to be circulated to, and read by, all involved.

In order to keep abreast of the latest developments, Merritt arranged for himself and his staff to maintain regular visits to relevant national and commercial institutions, such as the National Institute of Agricultural Engineering, the National Physical Laboratory and the Shell Oil Laboratory. Visits were also made to various types of farms - mixed, arable, dairy, even fruit - in different areas of the country to ascertain the differing needs of customers. This agenda was carried on after the war with trips to several continental agricultural shows as well.

LEFT:
Charles Hull with Fred Meadowcroft on the corner of Bent Ley Road, just around the corner from the Meltham plant. Hull became chief experimental engineer for David Brown Tractors in 1943 while Meadowcroft ran the field-test department.

VAK1 Production

After Merritt took over the responsibility for tractor design in 1941, one of his first initiatives was to create an alpha numeric scheme to identify the various types of David Brown tractor in production. These designations were used on engineering department documents and serial number plates to differentiate the agricultural models from the tractors for the Air Ministry. For example, the basic agricultural tractor with the TVO engine became VAK1 - standing for Vehicle Agricultural Kerosene No.1. For customers who preferred a petrol engine (such as for light industrial or municipal use, a VAG1 (Vehicle Agricultural Gasoline) variation was available with flat-head pistons instead of the dished-head

ABOVE: *David Brown tractors in production at Meltham during the Second World War. In 1941, Henry Merritt introduced the VAK1 designation to differentiate the agricultural model from the industrial types.*

type used on the TVO models. The Americanised fuel nomenclature reflected the terminology used by the War Office.

Further designations were used to differentiate between the tractors with and without power lift. The drawbar tractor was known as VAK1/100, while the model with the U36 hydraulic unit was designated VAK1/112. Merritt extended the scheme to cover all model variants and engine options either in production or planned for the future. When tracked machines were eventually introduced, the 'V' designation was reserved for all wheeled vehicles.

Apart from restricting output, the only othe immediate effect the war had on tracto production was to push up costs. By May 194 the list price of the basic VAK1 tractor on ste wheels had risen by nearly 50 percent (in line wit most other manufactured goods and commoditie of the time) to £275.

Things were much different within a couple years as shortages of raw materials – particular steel, iron castings, bearings and precision chair - began to take effect. Shipping losses during th Battle of the Atlantic and the Japanese conques in the Far East further exacerbated the situatio

LEFT:
A consignment of David Brown VAK1 tractors leaving Meltham in about 1943. The lorry is a 1935 AEC Monarch.

affecting supplies of rubber and the resins for plastic and Bakelite. This led to severe cutbacks in pneumatic tyre and electrical component production, and austerity was the order of the day.

Even tighter restrictions were placed on agricultural engineering by the Farm Machinery (Control of Manufacture and Supply) Order, issued on 12 May 1941 by the Minister of

LEFT:
During the war, several changes were made to the VAK1 in answer to material shortages, such as the cast-iron grille being replaced by one made from perforated steel. This 1943 tractor would have originally appeared as an austerity model on steel wheels.

ABOVE: *This 1943 model VAK1 was delivered new to A. L. Palmer of Lower Hinton Farm, Martock in Somerset. Due to wartime shortages, it was finished in brown paint rather than the usual hunting pink. It is seen harvesting in 1945 with an Allis Chalmers All-Crop 60 combine. The driver, Nigel Palmer, became David Brown's chief demonstrator.*

Agriculture. Under these emergency powers, which regulated types and quantities of machinery made or supplied, no machinery manufacturer or dealer could operate without a licence and every sale had to be authorised. A panel of consultants, which included Bert Jenkins, was also set up to advise the Ministry of Agriculture on production engineering matters.

David Brown managed to negotiate the continuance of its licence to build 1,000 agricultural tractors a year by promising to supply machines to the War Office. As farm machinery manufacturing was deemed to be non-essential work, material supplies remained difficult and by 1943 the VAK1 tractor had been re-launched as a wartime 'Austerity' or 'Utility' model with steel wheels being the only option. The Scintilla Vertex magneto ha[s] been largely superseded by a Lucas 4VRA type, an[d] coil ignition, electric start and pneumatic tyres we[re] not available. The cast radiator grille was replace[d] by a perforated steel one - often referred to as th[e] 'bullet-hole' grille - to save on materials. A 'wartim[e] emergency' water pump with plain bearings wa[s] also fitted during this period.

Since its introduction the production VAK1 ha[s] been finished in hunting pink with the whee[ls] painted the same colour as the rest of the tracto[r] and the David Brown name prominent on the sid[e] of the engine panels. Wartime shortages affecte[d] paint allocations and it became increasingly difficu[lt] to get the correct shade of red, which meant th[at] for several months the tractors were delivered what has been described as a dirty brown or a khaki colour. But most people (including customers who bought VAK1 tractors during the war) recall the paint as being similar to a red lead-oxide primer.

Work was still continuing on the new combined inlet and exhaust manifold, but as an interim measure an economy baffle was introduced in June 1943. Designed to reflect heat from the exhaust to the inlet manifold and allow a more rapid changeover from petrol to TVO, it was fitted to all tractors built from serial No. 4767. Owners of earlier tractors could also obtain the baffle from the factory by sending a donation of one shilling to the Red Cross Agriculture Fund.

The same factors that were causing material shortages were also affecting food supplies, and British agriculture was never more important. The degree of modernisation that had been forced

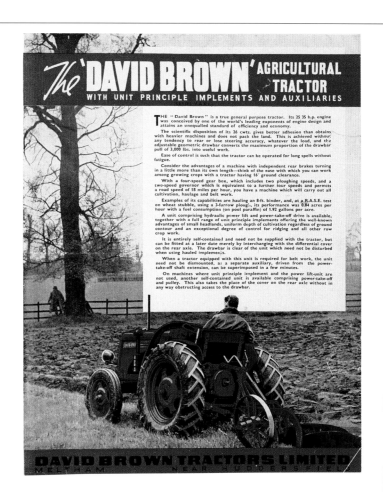

LEFT:
A 1943 sales leaflet for the David Brown VAK1 tractor.

BELOW:
The Box Mill at Meltham Mills housed both the parts and service departments for the tractor division during the latter years of the war. The building had originally been used for making cardboard boxes in the days of cotton production.

RIGHT:
In 1943, David Brown introduced a reconditioning service for its tractors. The overhauls were carried out by the service department at Meltham. The workshops were located on the ground floor of the Box Mill.

on farming since the outbreak of war meant that the demand for tractors was exceeding supply and those machines that were already out on the land had to be kept working. David Brown, therefore, introduced a reconditioning service at its Meltham plant.

The tractor overhauls were carried out by the servicing department, which worked in close association with the spares department. George Reekie had taken over as service manager after Bill Wallis had left to join Rolls-Royce on ordnance work, and Len Craven was now stores foreman. Reekie was a well-educated and somewhat enigmatic character whose influence seemed to extend as far as policy and board level, and he was often consulted on issues far beyond just service matters.

The combined tractor overhaul and spares department had been moved into the old Clock Mill in Easter 1941. As the scope of the tractor variants and the range of implements increased, the first floor of the mill (it had been nearly derelict before the service and spares department had moved in and was finally demolished in 1959) began to groan under the weight. This prompted the stores to be moved to a two-storey stone building, known as the Box Mill (originally used for manufacturing cardboard boxes), on the lower end of the Meltham Mills site near the cricket pitch.

The service workshop was located on the ground floor of the Box Mill. It was equipped with all the tools, drilling and valve-surfacing machines required for most major overhauls. The tractors were stripped on a makeshift assembly line with overhead gantries for

RIGHT:
The spares department for the tractor division at Meltham was housed on the upper floor of the Box Mill. The parts bins and storage racking were systematically arranged according to the tractor's assembly sequence.

removing and handling the major units and sub-assemblies.

An average of 120 man-hours was spent on each overhaul. All the components, down to even the last nut and bolt, were examined and then reconditioned or rebuilt as necessary. Any precision machining required was done in Meltham's main machine shop. As part of the final inspection, every engine was test-run on a dynamometer and had to be within 5 percent of its original power output before the tractor was passed for return to its owner.

The spares department was housed on the first floor of the Box Mill and relied heavily on female staff during the wartime period. The parts bins were arranged systematically according to the tractor's assembly sequence. Components from the stores were delivered to the assembly and feeder lines by a special trailer, which could be hoisted between the floors of the mill or turned into a handcart and wheeled around the bins and racking. Eventually the stores spilled out into some nearby old stone cottages originally built to house the mill workers.

Through 1943 and 1944, new equipment was developed and added to what David Brown called its range of 'unit principle' implements. First to appear in early 1943 was a new version of the three-row ridger with twin depth-wheel assemblies clamped to the angle-bar frame in front of the two outside ridging bodies. The bodies could be mounted in line or staggered.

In February 1944, the company demonstrated two new ploughs at Chaddleworth Farm, near Newbury in Berkshire. The first of these was a three-furrow version of the mounted ploughs already in production. Fitted with 10 in. general-

BELOW:
Designed by A. E. Smith, the David Brown potato spinner was developed in 1943 and launched the following year. It was driven from the tractor's power take-off and cost £56.

purpose bodies, it was launched as the PU4 model. The three-furrow made a good impression at the demonstration, but the same could not be said for the other plough, which was a strange two-furrow implement designed by local manufacturers, T. Baker & Sons (Compton) Ltd, to suit the David Brown tractor. It had a 6 in. furrow at the front and a variable-pitch 10 in. body at the rear. Although built for opening up virgin land, it proved almost unworkable at the demonstration and seems to have disappeared without trace.

A. E. Smith also developed a potato spinner for the VAK1. This machine had been in the

LEFT:
The potato spinner in the workshops of David Brown dealers, Belton Bros & Drury of Eastoft. The spinner had been brought in for modifications while Fred Meadowcroft was conducting field trials in Lincolnshire. Note the trailing plough in the foreground.

pipeline since Ferguson-Brown days in September 1938. It was mounted on the three-point linkage and driven by the tractor's power take-off. While field testing the implement in Lincolnshire, Fred Meadowcroft found that the digging arm, which was behind and in line with the top spinner, tended to bung up with haulm.

Using the blacksmith shop belonging to the local David Brown dealership, Belton Bros. & Drury of Eastoft, near Scunthorpe, Meadowcroft modified and repositioned the tubular-steel arm away from the spinner, thus preventing any further problem. He also fitted solid rubber tubes to the fingers of the spinner to avoid bruising the potatoes. Another field modification saw him strengthen the arm by welding ½ in. tube around it. It was only meant to be a makeshift field repair, but the designers copied the modification, drew it on to the plans and put it into production.

The spinner, designated DP2, was launched at a demonstration and exhibition staged at Eassie in Angus by the Highland and Agricultural Society of Scotland in April 1944. It proved to be such a popular machine that demand outstripped supply to the extent that the whole of 1945's allocation had sold out by the February of that year. It was not until September 1946 that the spinner, which cost £56, was available for almost instant delivery.

David Brown took the opportunity to introduce its new U28 rowcrop rear wheels at the Highland demonstration in 1944. Designed for general inter-row cultivation, the new rear wheels were made from narrow channel-section steel and the rims were only 2 in. wide. Possibly of greater interest at the demonstration was a clutch-release device for the VAK1, which had been invented by David Brown's main agents for the south-west of Scotland, McNeill Tractors & Equipment of Glasgow.

The McNeill clutch-release consisted of a contracting top-link connected by a wire Bowden cable to a trip mechanism on the hand-clutch control. The mechanism was demonstrated on a VAK1 working with a ridger. If the implement encountered an obstruction, the clutch was automatically disengaged to stop the tractor before any damage was caused.

The top-link consisted of a plunger and shaft held in tension by a coil spring. When the top-link contracted under load, four steel balls pressed against a cone to tension the Bowden cable. It was a simple device that was well received as an answer to the constant problem of cultivating stony land in Scotland.

David Brown's engineers were impressed with the McNeill device, particularly as safety considerations were becoming paramount in the design of agricultural machinery, and decided to secure the rights to the invention for production. Field trials highlighted the need for a few modifications, which were made by Fred Meadowcroft and included increasing the number of steel balls in the mechanism from four to nine. He also provided a Warwick screw to alter the pitch of the implements, thus introducing the turnbuckle top-link which was patented as another David Brown 'first'. The device, priced at £6 6s, was launched as the U37 overload release in November 1944.

At the same time, David Brown improved the diagonal link arrangement with the introduction of the U35 manual width-control hitch. Consisting of a winding lever that bolted to the right-hand lower link arm, it allowed later control of the plough from the driver's

BELOW:
A diagram of the David Brown U37 overload release mechanism, which was launched in 1944. When the top link contracted under load, it tensioned the Bowden cable, releasing the latch holding the hand-clutch lever to disengage the drive.

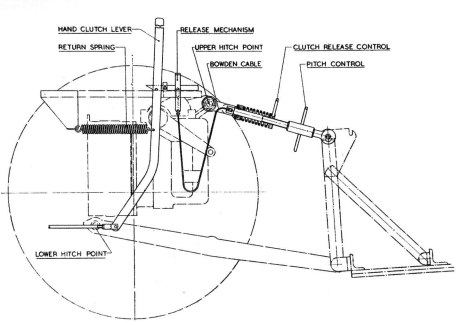

seat, even while the implement was in work. The mechanism came as standard with the latest power lifts or could be ordered separately by owners of the earlier hydraulic units for £4 7s 6d.

Other improvements to the David Brown tractor were in the pipeline and a revised model, launched in April 1945, marked the end of the VAK1 production, which totalled some 5,350 units. This figure masks the full extent of Meltham's wartime tractor production, which is revealed by comparing the number of VAK1s made to the number of tractor engines produced during the same period. The latter figure is greater by nearly 2,000 units and accounts for all the industrial models built for the Air Ministry and other branches of the armed services.

LEFT:
The U35 manual width-control hitch, which consisted of a winding lever bolted to the left-hand lower link-arm, was launched in 1944. This is the later U36A power lift with the adjustable top-link as fitted to the VAK1/A tractor.

David Brown's wartime tractor production was later summarised by Henry Merritt in a report (given to the Nuffield Organisation in 1947) describing Meltham's activities between 1942 and 1944. The report stated that average production was thirty-five tractors a week, accounted for by twenty-five agricultural models and ten Air

BELOW:
A wartime David Brown tractor on belt-pulley work driving a threshing drum. The VAK1 model was produced at the rate of about twenty-five tractors per week throughout the war.

RIGHT:
A 1942 David Brown VAK1 tractor at work. Production of the VAK1 model ended in April 1945 after some 5,350 had been built. This tractor features the VAK1/A engine modifications, which were available under a service exchange scheme after the improved model was released.

Ministry machines. This required a floor space of 30-40,000 sq ft with 133 machine tools (including twelve specialist tools for machining castings) and a staff of 350 personnel. The direct labour cost of each tractor was £18 and approximately 85 percent of the make-up of each tractor came from outside suppliers.

Air Ministry Tractors

Throughout the conflict, David Brown's agricultural tractor production was continually under threat from government officials who would have preferred to see Meltham's entire resources turned over to the war effort. It was often only the unstinting support of Mr David Brown's friends in high places, including Lord Beaverbrook and Lord

RIGHT:
A Ministry of Supply exhibition, held in the concert room at Meltham Mills in 1943, showing the range of military products made by David Brown Tractors Ltd. The defence contracts would have threatened tractor production had the company not had the support of several government ministers.

LEFT:
Thoughts of industrial models had existed since the inception of the David Brown tractor. This 'highway' or semi-industrial model, based on one of the original pre-production tractors, appeared at Park Works in mid-1939. Fitted with electric starting, lighting and a horn, it was designed for municipal use.

Brabazon of Tara that kept the VAK1 on the line. Brown also found an unexpected ally in Robert Spear Hudson, whom Churchill appointed as Minister of Agriculture in 1940. Hudson's almost authoritarian and often pedantic rule over the countryside did not always endear him to farmers, but he was a great proponent of the campaign for mechanisation. His continual exhortations to the Ministry of Supply for increased farm machinery production undoubtedly helped David Brown maintain its position as a tractor manufacturer, which Hudson argued should be encouraged to meet the needs of post-war agriculture.

The true saviour of the David Brown tractor was, however, the offshoot production of towing vehicles for the Air Ministry. By 1941, the aero-gear and tank gearbox production at Meltham was of such vital national importance that the agricultural activities were under real threat. After President Roosevelt signed the Lend-Lease Bill in March, allowing Britain to obtain arms, vehicles and supplies from the USA with payment deferred until after war, the dissenters at the Ministry of Supply suggested that there was now no excuse for allowing the David Brown tractor to remain in production; American machines were becoming available and shipments of foodstuffs were being stepped up.

Lord Beaverbrook thought otherwise and outlined the need for industrial variants of the David Brown tractor for use on the RAF's airfields, arguing that similar machines were not being imported from the USA. Furthermore, too much reliance on imports would be a dangerous move while U-boat attacks on merchant shipping continued. The argument was based on good sense and David Brown tractor production was reprieved.

David Brown had been developing semi-industrial versions of its tractors for some time, but as yet had put nothing into production. One of the original pre-production tractors built at Park Works had even been turned into a 'highway' model for municipal use as early as mid-1939 but was never launched. A semi-industrial version of the VAG1 had also reached prototype form, but was not really suitable for the requirements of the RAF, which wanted something more specialised.

The development of David Brown tractors for the Air Ministry was inextricably linked to the expansion and changes affecting Britain's military airfields. After the First World War, little credibility had been given to air defence and, as unbelievable as it may sound, in the early 1920s the country was protected by one lone RAF fighter squadron and just four Vickers Vimy bombers. There were even parliamentary attempts to disband the RAF as the airfields fell into neglect.

During the 1930s, Britain reluctantly began to boost its defences to counter the growing threat of Nazi Germany and the increasing strength of its military airforce, the Luftwaffe. The summer of

1934 saw the beginning of the so-called 'Expansion Scheme' for Britain's military airfields. This expansion period lasted into the early years of the Second World War, gaining greater impetus after the Munich crisis of 1938. Disused airfields were reactivated and new bases were commissioned and constructed with purpose-built brick buildings and hangers replacing the earlier temporary structures.

Most of the new aerodromes were built on agricultural land that was levelled and grassed. Tractors were needed to mow grass runways, clear snow and move aircraft. Crawlers were chosen because of their extra traction and ability to work on all ground conditions. The only British manufacturer of light crawlers was Roadless Traction Ltd of Hounslow in Middlesex, a firm that built tracked conversions of other manufacturers machines. The first tractor supplied to the Air Ministry, a Roadless Rushton delivered to Biggin Hill in 1932, was followed by further orders for Roadless Fordson models in 1936. Problems with the crawlers pitching while mowing led to the tractors being fitted with a fore-carriage arrangement from 1938. A front-mounted Hesford winch was also mounted on the fore-carriage platform.

The rapid expansion of the military airfield construction programme during the early years of the war left the Air Ministry seriously short of tractors and Roadless, a relatively small engineering concern, struggling to meet the demand. Consequently, in 1940 David Brown was approached to build a tracklayer along similar lines to the Roadless crawler and meet the Ministry's requirements for a machine that could be used for both aerodrome construction and aircraft towing.

The David Brown crawler, put together by a team working under Albert Kersey

BELOW:
This semi-industrial tractor, based on the VAG1 model with heavy-duty wheel centres, was being developed at Meltham in 1940, but probably never went beyond the prototype stage due to the pressures of wartime production.

RIGHT:
RAF observers attend a drawbar test of the prototype Air Ministry crawler. David Brown developed the tractor in 1940 to meet the ministry's requirements for a machine for airfield use. Demonstrator Bill Harrison (in black jacket and white overalls) can be seen in the centre of the group. The location is the Bar House field, on the corner where Meltham Mills Road joins the Huddersfield Road.

ABOVE:
The prototype David Brown Air Ministry crawler on trial on moorland above Meltham Mills. The steel tracks were supplied by Roadless.

was not so much designed as assembled from what was available. It was basically an agricultural tractor skid-unit with a petrol engine dropped on to a sub-frame that carried the track-rollers, front idlers and a front-mounted Hesford winch. Few modifications were made other than alterations to the gearbox ratios to give lower speeds for towing. Top gear was also blanked-off as high speeds would only cause unnecessary damage to the tractor. Roadless supplied the tracks; not the silent-running rubber-jointed type as used on its own crawlers but a heavy-duty girder type of a conventional pin and bush construction.

The crawler was steered by its independent brakes, which were controlled by two hand levers. There was no foot brake and braking on an incline or at speed was carried out by the slightly precarious method of pulling simultaneously on both hand levers. The steering levers were also fitted with retaining pawls to act as a parking brake.

The winch was the same model as fitted to the Roadless tractors and was supplied by C. M. Hesford of Ormskirk in Lancashire. A sprag, which acted as a land anchor when the winch was in use, was mounted beneath the crawler and was raised or lowered by a lever to the right of the driver's seat. The winch was driven by the tractor's rear power take-off and the drive was taken via dog-clutch to a Cardan-shaft that extended the full length of the crawler. Control levers for the clutch and brake for the winch were mounted on the nearside track guard within easy reach of the driver.

The tractor was usually referred to as the Air Ministry crawler. Documents show that it was provisionally designated AC1 (Air Crawler One) with its serial numbers prefixed by the letter C. A couple of prototypes were prepared for trials in May 1940, and one of these was loaned to the Royal Engineers for use on the construction of coastal defences. However, the crawler was pressed into production almost immediately, before any trials had been completed, because of the urgent need for machines to assist in the preparation of roadways and airbases for the campaign in the Western Desert.

The first batch of crawlers (serial numbers C1 to C150) was delivered between June and December 1940. Most of these were diverted from the Air Ministry to the Royal Engineers and shipped out to North Africa. At least one of these tractors had an interesting campaign; after being captured by the Italian army, it was liberated by the British Eighth Army after El Alamein in October 1942. It then passed into the hands of 889 Naval Air Squadron, a Fleet Air Arm unit seconded to the RAF, which used it for establishing bases on the Egyptian coast for shipping protection and canal-zone defence until February 1943.

ABOVE:
One of the first batch of David Brown Air Ministry crawlers. Built between June and December 1940 and fitted with a front-mounted Hesford winch, most of these machines were diverted to the Royal Engineers and saw service in North Africa. Note the Cardan-shaft drive to the winch.

Initial field trials, which included timber extraction and clearing snow from the roads around Meltham through the 1941/42 winter, highlighted a number of weaknesses with the crawler, not least in its Hesford winch, which suffered from brittle castings and weak bearings. Its cast bevel-reduction and spur gears were also not up to the job. David Brown decided to develop its own version of the winch with better castings, stronger bearings and machine-cut gears. Charles Hull carried out most of the work and the new winch had a 10,000 lb pull and held 100 ft of cable. Its clutch-control was mounted on the winch while the brake-control lever remained on the track-guard. Fred Meadowcroft also added two bronze pulleys to the fairleads on the winch frame to avoid problems with the rope knotting.

While testing one of the crawlers during the winter, Meadowcroft found that soil tended to build up inside the rear drive-sprocket, which was made from two solid plates. Snow also created a problem and would pack inside the sprocket and break the tracks. His solution was to replace the outer plate on the sprocket with an open ring so that the soil or snow would fall through.

Further alterations saw the two-speed governor control rod for the engine deleted from the specification. The brakes were also modified and a single brake pedal that operated on both independent brakes was provided as a safety measure. In addition, the Air Ministry wanted a vacuum system for trailer braking. This was taken off the inlet manifold and operated through a Clayton Dewandre reaction valve which connected the foot-brake linkage to provide single-line braking.

The modifications, including the new David Brown winch, were incorporated into the second and final batch of Air Ministry crawlers (serial numbers C151 to C551), which were built from around January to October 1941. These were supplied almost entirely to the Air Ministry for use on airfields.

A few of the tractors were employed by the RAF's No. 50 MU (Maintenance Unit). This unit, based in Oxford and staffed by Morris Motors personnel, offered a twenty-four hour service recovering crashed aircraft from bombed aerodromes during the Battle of Britain. The wrecked aeroplanes were stored on a 100 acre site near Oxford and stripped to provide much needed parts and materials for the aircraft industry. It was part of a nationwide salvage drive for aluminium, initiated by Lord Beaverbrook and promoted by the slogan: 'We will turn your pots and pans into Spitfires, Hurricanes, Blenheims and Wellingtons.'

LEFT:
This David Brown crawler is seen in Egypt in the hands of 889 Naval Air Squadron after being captured by the Italians and then liberated by the British Eighth Army at El Alamein in 1942.

LEFT:
A David Brown crawler is used for snow ploughing the roads around Meltham. While testing in the winter, Fred Meadowcroft found that snow tended to pack into the solid back sprocket, which was eventually replaced by an open ring.

No more than the two batches of Air Ministry crawlers, totalling some 550 odd machines, were built before assembly was halted because the changing nature of Britain's aerodromes dictated that a different type of vehicle was needed. From 1940, most of the new airfields had been constructed with concrete runways and

ABOVE:
A David Brown Air Ministry crawler outside the Aero Block at Meltham with a wrecked German Messerschmitt Bf 109 fighter. Several of the crawlers were used by an RAF maintenance unit for recovering crashed aircraft.

perimeter roads. The crawlers damaged the concrete and were uncomfortable to drive on the hard surfaces. They were also difficult to manoeuvre and were too slow when speed was of the essence in preparing aircraft for operations.

The RAF had a considerable number of wheeled Fordson tractors, some of which had been in service since 1938. While these were capable of moving or servicing the smaller twin-engined bombers, most were confined to Fighter Command and used for refuelling duties. The Air Ministry's latest requirement was for heavy wheeled tractors to handle the new breed of four-engined bombers - the Short Stirling, the Handley Page Halifax and the Avro Manchester (the precursor to the Lancaster) - that were coming into service as the RAF turned from defensive to offensive operations.

It was not just a question of moving the bombers on the apron. The tractors also had to be able to handle the considerable logistics of preparing the big planes for action - towing fuel and oil bowsers and trolleys for 'bombing up' the aircraft for raids over enemy territory. To give some idea of the logistics involved, some of the larger raids, where Bomber Command might put up over 1,000 aircraft on a single night, would require 2 million gallons of petrol, 70,000 gallons of oil and 5,000 gallons of coolant. On top of this, 4,500 tons of bombs, 10 million rounds of ammunition, 15 million litres of oxygen, 8,000 pints of coffee and 6,000 lb of food were distributed between the aircraft.

For obvious safety reasons, the bomb and ammunition stores were normally sited in the remotest areas of the airfield, well away from the main runway, hangers and living quarters. The Lancaster, for example, had an average bomb load of around 12,000 lb and the tractors had to have sufficient drawbar pull to be able to handle trains of up to six bomb trolleys to avoid too many time-consuming trips across the airfield. They also had to be capable of salvage operations and to be able to clear the runway at a moment's notice.

On some of the larger operations, up to twenty-five Lancaster bombers might be lined up at dispersal ready for take-off at one time. If one plane went 'u/s', it could compromise the whole raid, and so towing vehicles were always on hand to haul them off the runway as quickly as possible. This is why the aircraft tractors were so important to the war effort.

David Brown was delighted, if a little surprised, to be awarded the sole contract for the design and production of the heavy wheeled-tractors. The Air Ministry's brief was for a vehicle 'built primarily for the towing of heavy aircraft, bomb carriers, tankers and other mobile equipment associated with Bomber and Fighter Command operations. The general dimensions, weight distribution and tyre equipment must be designed to give maximum adhesion.'

The contract was worth £1½ million and would be a considerable undertaking for a relatively small concern. With no prospect of payment until after delivery, it represented such a significant drain on finances that Mr David Brown had to approach his bank for backing before he dared to accept the contract.

Work on designing the Air Ministry wheeled

BELOW:
The prototype Air Ministry wheeled tractor outside the gates to Meltham Hall. Introduced in 1941, these heavy towing vehicles were designed to meet the RAF's latest requirements for moving heavy bombers.

RIGHT:
David Brown aircraft-towing tractors on the assembly line at Meltham. The steel channel and supports that made up the sub-frame can be clearly seen with the RAF standard towing hook in place.

tractors began in 1941 with Bill McCaw returning as chief draughtsman in January to take charge of the project. The machine was designed and built almost from scratch in the main tractor machine shop without any development work being carried out.

The vehicle was based on an agricultural skid-unit with a petrol engine and a four-speed gearbox. The skid-unit was mounted on a sub-frame made from steel channel and the castings were supported in three places - at the front and on the two final-drive reduction units. The final drives incorporated lower ratios than those fitted to the agricultural tractor and were housed in larger rectangular casings. The sub-frame, which also supported the front axle, extended rearwards to provide a rigid mounting for the winch, heavy-duty sprag and an RAF standard towing hook. The winch was driven from the power take-off via a chain, bevel-reduction and spur gears.

The cast rear wheels had integral brake drums and were fitted with Dunlop 'Ground-Grip' tyres. Ballast weights were also bolted to the sub-frame underneath the centre of the tractor to provide maximum traction. A single brake pedal applied both rear brakes simultaneously and a foot-throttle was also provided. Full-width fenders made from heavy-gauge steel completed the specification and gave the tractor a purposeful appearance.

In wartime RAF nomenclature, it was known as the David Brown 4 x 2 Light Tractor Type AW500. The use of the military 'Light' designation is typically inappropriate considering the machine weighed 3 ton 17½ cwt and had a drawbar pull in first gear of 4,500 lb. Some of the prototypes and the very first few tractors used for trials either had Hesford winches or no winch at all. Initial trials were conducted in Yorkshire at RAF Leeming, where the tractors were used to tow Halifax bombers.

Production models, fitted with David Brown winches and vacuum trailer braking, began to enter service at the end of 1941. Their appearance coincided with the adoption of Henry Merritt's alpha numeric classifications. Under this scheme the Air Ministry tractors were given the VIG1 (Vehicle Industrial Gasoline) designation. The AW500 was reclassified as the VIG1/100 but continued to use the AW prefix on its serial numbers. However, RAF documentation of the wartime period continued to make a distinction between the AW500 and VIG1/100, indicating that the latter model had an improved winch and vacuum braking system. There will always be confusion over the designations for the Air Ministry tractors because the RAF would occasionally alter them to suit local requirements or to cover modifications carried out while the vehicles were in service.

The first batch of Air Ministry wheeled tractors (serial numbers AW100 to AW750) was built between October 1941 and December 1942 and accounted for some 650 vehicles. To accommodate the extra production, the tractor assembly shop was reorganised on two levels with the agricultural machines being built on the lower floor and the Air Ministry vehicles assembled on the first floor.

From mid-1942, a limited number of civilian versions of the same machine was released to firms and contractors with the necessary Ministry of Supply permits and engaged on essential war production, such as aircraft and munitions output. These tractors were identified by an IND rather than an AW prefix to their serial numbers.

Earlier in the year, interested parties including government officials had been invited to Meltham for a demonstration of the industrial tractor and, in particular, its David Brown winch. The demonstration, which was staged in the yard below the Aero Block, involved using the winch to haul the tractor up on its own sprag until it was suspended nearly four feet off the ground. It was an impressive display both of the power of the winch and the driver's nerve!

Most of the crawlers supplied to the Air Ministry were recalled to Meltham to be converted into wheeled tractors, emerging as almost identical machines to the VIG1/100 except for different third-gear ratios. The logistics of the undertaking turned out to be something of a

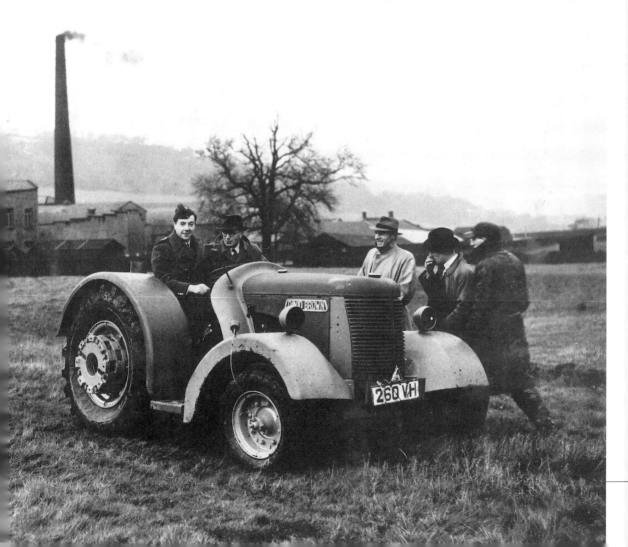

LEFT:
One of the prototype Air Ministry tractors on test in 1941 at the Bar House. Note the extra weights added to the rear wheels. Further ballast weights were bolted to the sub-frame.

ABOVE:
The drawbar pull of a prototype wheeled towing tractor is tested against one of the Air Ministry crawlers. Many of the crawlers supplied to the RAF were later recalled to Meltham and converted into wheeled tractors.

nightmare for David Brown. Hundreds of tractors waiting to be converted had to be stored at Scarr Bottom and other satellite depots as far afield as Penistone. The converted crawlers were initially designated AW100 (by the RAF) and their serial numbers were prefixed AWC.

However, the Air Ministry's requirements were due to change yet again. The heavy bombardment and strafing of British airfields in the south of England between August and October 1940 had been a great cause for concern and had led to a policy rethink regarding the deployment of aircraft. The heavy bombers sitting on concrete runways on the larger airfields were judged to be sitting targets for the German fighters. Therefore, during 1941 the decision was made to disperse the aircraft either to the grass perimeter of the base or to remote second-line stations, known as SLGs (Satellite Landing Grounds) with grass runways and few buildings or facilities.

This new dispersal policy caused problems of its own when periods of heavy rain led to the aircraft becoming bogged down on the grass. The David Brown tractor was ideal for winching them out but sometimes struggled on a direct pull. The undue strain of trying to move a heavy bomber that had sunk into the grass sometimes caused the dry clutch to slip or else resulted in a snatched pull, particularly if an inexperienced driver was at the wheel, with the risk of damage to a precious aircraft undercarriage. The wet clutch (running in oil) that was fitted to the Fordson was more forgiving, but the Dagenham-built machines had neither the power nor the weight to move the big bombers, and crawlers had already been dismissed because they cut up the grass runways. A drastic solution was required and David Brown came up with the answer.

Bill McCaw had been recalled to David Brown because of his experience in hydraulics and because the company was considering fitting a torque converter transmission to its Air Ministry tractors. A suitable torque converter was under development with Brockhouse Engineering, who had taken over the old Vulcan lorry works at Crossens near Southport. McCaw liaised with Brockhouse's design engineer, Andy Anderson and the project developed into a joint undertaking with the Ministry of Supply.

A torque converter is a fluid drive-unit that acts, within a certain range, as an infinitely variable transmission that adjusts itself automatically to the load. An integral pump circulates hydraulic fluid to pressurise a driven turbine, thus multiplying the input torque from the engine. Maximum torque multiplication occurs at high loads and low speeds.

The model chosen for the David Brown tractor was known as the Brockhouse 12½ in. Turbo Transmitter, a unit that gave an automatic torque magnification of 3.5 to 1. It was ideal for a steady pull at low speeds and was particularly helpful for starting a heavy load from rest without snatch or shock. It maximised the static adhesion between the tyres and the ground and eliminated wheel slip.

A prototype vehicle fitted with the turbo transmitter was loaned to the Ministry of Supply and used for trials at Oakington aerodrome near Cambridge during 1942. It proved easily capable of moving the heavy bombers on grass without the risk of damage either to itself or the aircraft. It led to the development of a tractor based on the VIG1/100 with the Brockhouse unit installed in place of the clutch assembly between the engine and the normal four-speed gearbox.

By retaining the gearbox, the engine speed could be synchronised to the ground speed to avoid overloading or overheating the torque converter. A rocking brake, operated by a hand lever, arrested the motion of the turbo transmitter's output shaft to overcome oil drag so that the gears could be engaged. The drive to the winch also came through the fluid transmission, allowing the load to be smoothly and uniformly increased.

During trials of the new tractor, one of Bill McCaw's designers, Tommy Allen, put one of the front wheels against a concrete bollard to test the stalling torque of the turbo transmitter. The power of the torque converter was such that the front wheel unexpectedly rode vertically up the side of the bollard and dropped down the other side,

LEFT:
This photograph underlines the logistics of preparing one of the RAF's new breed of four-engined bombers, in this case a Short Stirling, for a wartime raid. The service equipment includes two David Brown towing tractors and an AEC Matador tanker.

RIGHT:
Production models of the David Brown aircraft-towing tractors, designated VIG1/100, entered service with the RAF at the end of 1941. This example is seen towing a Lockheed Ventura light-bomber in 1942.

BELOW:
A David Brown VIG/100 hauls a train of 2,000 lb bombs out of the ammunition dump. The tractor had a drawbar pull of 4,500 lb.

leaving the tractor's sub-frame wedged solid. It took most of a day to jack the machine up and extricate it from the bollard. The exercise was not repeated!

Following trials towing Lancaster bombers in the snow at RAF Waddington in Lincolnshire, the new tractor with the turbo transmitter was put into production in January 1943. It was introduced from serial number AW 751 onwards and was designated VIG1/462. Apart from the fluid transmission, it was identical to the VIG1/100 model, which it replaced. However, it did weigh slightly less at 3 ton 16½ cwt and had an increased drawbar pull in first gear of 5,400 lb. It became the standard aircraft towing tractor until the late 1940s and was supplied both to the Air

LEFT:
Logging with a David Brown industrial tractor. A few civilian versions of the Air Ministry machines were released to contractors involved in essential war work from mid-1942.

Ministry and to the Admiralty for use with the Fleet Air Arm on aircraft carriers.

Soon after the release of the VIG1/462 model in 1943, Bill McCaw, having never really seen eye to eye with his immediate superior, Henry Merritt, left the company to take up a position with Steels Engineering Products of Sunderland. Herbert Ashfield then took over the responsibility as chief draughtsman for both the tank gearbox and tractor divisions at Meltham.

LEFT:
An impressive demonstration of the power of the David Brown winch, staged at Meltham in 1942. The photograph shows the winch, which had a 10,000 lb pull, being used to pull the tractor up on its spag until the whole machine was suspended in mid-air. The winch had been designed by Charles Hull after earlier Hesford units had proved to be too weak.

After leaving David Brown, McCaw had a varied career. His new firm, Steels, was the parent company of Coles Cranes and was swamped with defence contracts for work ranging from snowploughs and anchors to heating and ventilation systems for military camps and warships. He spent the rest of the war developing electric vehicles at Steels' Crown Works and then, after a short spell with Rotary Hoes at East Hornden, joined Muir-Hill as chief designer to

LEFT:
Three rather battered VIG1/100 tractors and an Air Ministry crawler possibly returned to David Brown for reconditioning or overhaul. The photograph is taken in the old wood-yard at the top of Knowle Lane where timber was stored for bobbin making. Hundreds of crawler tractors awaiting conversion into wheeled machines were stored at satellite depots all over Yorkshire.

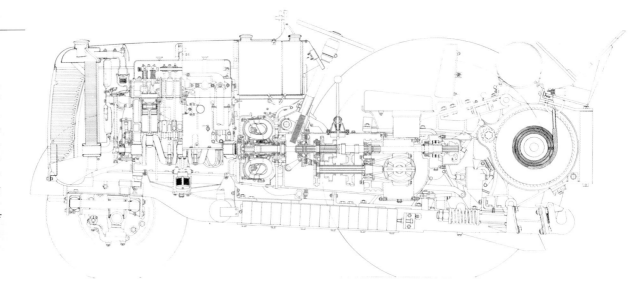

RIGHT:
A drawing showing the general arrangement of the new VIG1/462 towing tractor with the Brockhouse torque converter. It was developed during 1942 to meet the latest requirements of the Air Ministry.

RIGHT:
The Brockhouse 12½ in. Turbo Transmitter fitted to an experimental David Brown VIG1/462 tractor. The torque converter, in simple terms a turbine pressurised by a hydraulic pump, was installed in place of the clutch assembly between the engine and gearbox.

investigate the use of hydraulics in dumpers and loading shovels.

The contracts for the Air Ministry tractors only served to add to Meltham's workload during the Second World War. The factory would often work around the clock with twelve-hour day and night shifts in operation. At times of utmost urgency, it was not unknown for a seven-day eighty-four-hour working week having to be implemented.

The production of the David Brown towing tractors for the

RIGHT:
The prototype David Brown VIG1/462 tractor with the turbo transmitter transmission. The photograph gives a good view of the rear-mounted David Brown winch and the heavy-duty land-anchor or sprag. The vacuum reservoir for the air-braking system can also be seen behind the sprag

armed services came under the direction of the Ministry of Supply acting on behalf of the Air Ministry, the Admiralty and the War Office. Defence contracts by their very nature were subject to stringent and comprehensive regulations and generated reams of documentation. To ensure that the strict guidelines were adhered to, resident government inspectors remained on site at the plant to represent the interests of the ministries involved and liaise with Meltham's various manufacturing departments.

The tank gearbox division came under the aegis of the Inspectorate of Fighting Vehicles while the aircraft towing tractors became the responsibility of the Admiralty Inspectorate Department. The first resident AID inspector from 1941 was George Booth. As the work with towing tractors expanded, extra inspectors were brought in under the supervision of the senior AID inspector, Sidney James, who liaised directly with Herbert Ashfield.

It has to be said that the inspectors' preoccupation with red tape did not always endear them to the David Brown men, whose main desire was to get the job done as quickly and efficiently as possible. The odd confrontation usually resulted in an unresolved truce as the two factions settled down into an uneasy working relationship. As one former Meltham employee put it: 'It was the case of the unshakeable force of a government official meeting the unmovable object of a bluff Yorkshireman.'

Before the AID officials would sign off any of the aircraft-towing vehicles for delivery, each tractor was given a strenuous twenty-four mile

ABOVE:
This VIG1/462 tractor with the turbo transmitter was used for trials at RAF Waddington in January 1943, towing and servicing Lancaster bombers in the snow. The vehicle's fluid drive allowed the maximum pull to be achieved without snatch or shock.

LEFT:
A David Brown aircraft-towing tractor on test with one of the RAF's Karrier CK6 machinery trucks. The contracts with the Air Ministry increased Meltham's workload and the factory often worked around the clock to meet orders.

RIGHT:
Air Ministry tractors on test at Meltham. Each tractor was given a 24 mile road test with an RAF driver at the wheel before they were signed off for delivery by the resident Admiralty inspectors.

BELOW:
A David Brown tractor 'bombing up' Handley Page Hampden aircraft with shipping mines. David Brown supplied around 1,350 wheeled tractors to the Air Ministry during the Second World War.

road test, usually with an RAF driver at the wheel and an AID inspector sitting next to him. This was no real hardship because the VIG1/462 tractors had a decent turn of speed. Mr David Brown also liked to conduct his own test drives and occasionally took one of the Air Ministry machines out for a spin round Meltham Moor. Arthur Caudwell, one of the fitters from the experimental machine shop, remembers accompanying him on one such jaunt over to the Isle of Skye inn and being frightened out of his wits by his boss's driving!

Government contracts for the towing tractor came to an end with the cessation of hostilities. B December 1946, production figures for the A Ministry and industrial tractors were in the regio of 1,350 AW series and 450 IND series machines i addition to the original crawlers. It is the VIG1/46 model that is most fondly remembered by forme David Brown personnel. Herbert Ashfield referre to it as 'the tractor that could move the world'.

The DB4 Crawler

Unbelievable as it may sound, there were no medium-weight or heavy crawler tractors or bulldozers in production in Britain during the Second World War. For that matter, apart from the few David Brown Air Ministry crawlers and the Roadless machines, most of which were now supplied as half-tracks with the fore-carriage arrangements, there had not really been any crawlers of any sort made in the country since the start of the conflict. John Fowler of Leeds had built a range of diesel crawlers before the war, but the company was annexed by the Ministry of Supply after running into financial difficulties and turned over almost totally to tank production.

Now, more than ever, crawlers and bulldozers were needed for military and defence work, airfield construction and land reclamation. The War Office, the Ministry of Supply and the Ministry of Agriculture all had to rely on imported Caterpillar, Allis-Chalmers and Cleveland Cletrac machines shipped from the USA.

In America, the US Army engineering battalions had all but standardised on the rugged and reliable diesel Caterpillar tractors, using D4, D6, D7 and D8 models equipped with LeTourneau and LaPlante-Choate angledozer and bulldozer attachments. The British Army also liked the big Cats and imports of tractors, dozers and scrapers for military use had been steady since 1937. D7s with LaPlante-Choate bulldozers were used to shift the debris from London streets during the Blitz while the Royal Engineers used D4 tractors for clearing fire-zones and constructing coastal defences in the so-called 'Defence Areas' in southern and eastern England. The sappers found the Caterpillar D4, known as the 'M2 light tractor' in US Army nomenclature, the ideal machine for most of its general construction work. The problem was that, as the war progressed, supplies of Caterpillar equipment became more difficult to obtain.

Things came to a head once America entered the war after Pearl Harbor in December 1941. This led to nearly 90 percent of the machines made at Caterpillar's Peoria plant being earmarked for national defence (building ordnance depots, factories and airfields) and for military use in the war with Japan in the Pacific, Burma and the Far East. This left just over 10 percent of 'Cat' production to meet the requirements of the civilian customers and America's allies. On top of this, the U-boat blockade was being stepped up and extra shipping could not be spared to bring any more crawlers into Britain for the foreseeable future.

As the situation became critical, the Ministry of Supply looked into the possibilities of sourcing suitable machines at home. W. E. Bray of Isleworth

BELOW:
An artist's impression of the proposed David Brown crawler based on the Caterpillar D4. The company at Meltham had been approached by the Ministry of Supply to build such a machine for military use. The picture shows the tractor fitted with LaPlante-Choate equipment and what looks suspiciously like a Caterpillar engine.

in Middlesex had been building hydraulic bulldozers and angledozers since 1938 and had produced a batch of eighteen machines, mostly based on Roadless crawlers, for the War Office. The company had also assisted in the conversion of a number of obsolete tanks, including the Nuffield Crusader, into bulldozers but these machines were heavy, cumbersome and difficult to manoeuvre.

Trials in January 1941 with a Roadless crawler, based on a Case Model L tractor and equipped with a Bray bulldozer attachment, sadly came to nothing. The machine did not meet the

RIGHT:
This 2T model Caterpillar D4 was delivered to Meltham in 1941. David Brown's engineers completely stripped the tractor so that the draughtsmen could make drawings from the parts. The exercise was carried out with government approval.

BELOW:
The engine chosen for the new David Brown crawler was the four-cylinder Dorman 4DWD - a new high-speed direct-injection diesel designed for automotive use. Built in Stafford, it had a 5.4 litre capacity and used CAV/Bosch fuel-injection equipment.

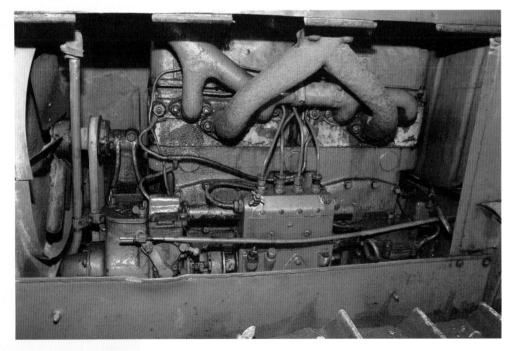

requirements of the Royal Engineers, who wanted something stronger with a diesel engine, and the American Case skid-unit would have been affected by the same supply difficulties as the Caterpillar tractors.

Thus, the Ministry of Supply approached its old colleagues at David Brown Tractors Ltd and asked the company if, with the government official sanction, it could make an exact copy of the Caterpillar D4 and treat it as an urgent order for almost immediate production. The first th engineers at Meltham knew about it was when D4 tractor was delivered to the experiment department and they were asked to strip it dow completely and thoroughly clean th parts so that the draughtsmen coul make drawings of them.

The prospect of making a copy of th engine was never considered. Tooling-u for engine production would have bee prohibitively expensive and too tim consuming. Furthermore, it might hav been pushing the boundaries of Caterpillar's goodwill, if indeed any suc agreement was in place. David Brown engineers cast around for a suitab power-plant and found one in the form a new direct-injection diesel engin recently introduced by Dorman.

For some time, W. H. Dorman & Co. Lt of Stafford had been building a range indirect-injection diesel engines, main for industrial use, with Ricardo rotar swirl combustion chambers. In Octob

1936, the company launched a new range of direct-injection power units. Although they used the same blocks with the same bore and stroke as the old Dorman-Ricardo range, the new engines had a different head with an open combustion chamber and employed CAV/Bosch fuel-injection equipment.

The new Dorman range of high-speed diesels consisted of two-, three- and six-cylinder units, primarily for industrial applications, as well as a four-cylinder motor suitable for transport duties. The latter engine developed up to 72 bhp at 2,000 rpm and was the unit chosen for the crawler because its power output, performance characteristics and general dimensions most closely matched those of the Caterpillar diesel.

The engine was known as the Dorman 4DWD. It had a 4½ in. (115 mm) bore and a 5⅛ in. (130 mm) stroke giving a cubic capacity of 329 cu in. (5,400 cc). Its dry cylinder liners were made from high-grade cast-iron, which had been chrome-hardened using the latest Listard van der Horst process. Special aluminium-alloy pistons and a five-bearing crankshaft completed the specification of what was a quite excellent and modern engine for the time.

The version of the engine chosen for the David Brown crawler was governed to 1,400 rpm, at which its output was rather conservatively rated at 45 bhp. When fitted to the tractor, this would equate to an equivalent drawbar horsepower of 38.5, which was about as close as you could get to the Cat D4's 35 dhp.

The new crawler was christened the David Brown DB4 - an apt if rather unoriginal identification. Confusingly, its model designation and serial number prefix, like the Air Ministry crawlers, was AC1. As the design work on the DB4 progressed, a power unit from one of the earlier Air Ministry crawlers, which was at Meltham awaiting conversion into a wheeled tractor, was fitted to the Caterpillar D4 chassis and running gear. This hybrid machine was then used for evaluation purposes.

LEFT:
David Brown's engineers used this hybrid machine for evaluation purposes. It consisted of the Caterpillar D4 chassis and running gear fitted with the early Air Ministry crawler power unit.

The David Brown Tractor Story

RIGHT:
The prototype David Brown DB4 crawler. It was very similar to the Caterpillar machine, but a number of features were altered to suit Meltham's production methods.

BELOW:
A rear view of the DB4 crawler. The tracks and running gear were almost identical to those of the Caterpillar, but most other parts were dimensionally different.

The DB4 was not an exact replica of the D4, as has often been stated. The running gear and tracks were almost identical, but most of the other components were dimensionally different from the Caterpillar parts and were most probably altered to suit Meltham's factory methods and tooling. David Brown also used bolts with BSF (British Standard Fine) threads while Caterpillar used American National Fine/Coarse with locking nuts. This later led to a few problems with the nuts and bolts on the Meltham-built machine vibrating loose and bits dropping off! It also meant that, although the DB4 had the same features as the D4, including the five-speed gearbox and the clutch and brake steering, very few parts were interchangeable. The David Brown tractor, of course, had electric starting with CAV electrical equipment while the Caterpillar relied on an auxiliary 'donkey' engine.

Components for the tracks were supplied by Tractor Spares Ltd of Willenhall in the West Midlands - a company specialising in replacement parts for Caterpillar tractors. Tractor Spares proprietor, Charles Weight, had bought up the entire spares belonging to the former Caterpillar agents, Tractor Traders Ltd, in 1936. During th

LEFT:
The production line for the DB4 crawlers at Meltham Mills. Tracks were supplied by Tractor Spares Ltd from the West Midlands.

Second World War, he carried out tractor repairs for the Ministry of Supply and, through a subsidiary company, acted as a contractor for aerodrome construction.

One of the first batches of DB4 crawlers was built under strict secrecy and with some urgency during the summer of 1942. The tractors were needed for a raid on the French coast of Dieppe, probably to construct defensive earthworks after the landing. Some of the David Brown personnel worked a non-stop thirty-six hour shift to prepare the crawlers for the raid, which took place on

LEFT:
The DB4 crawler was supplied to the Royal Engineers for military use and most were fitted with LaPlante-Choate bulldozing equipment. The tractors were involved in both the Dieppe raid and the Normandy landings.

RIGHT:
The DB4 was made in small batches up to 1945. This machine is seen clearing land ready for an extension to the Meltham Mills factory. The official production figure for these crawlers was 110.

19 August and sadly proved to be a disaster with nearly half of the 6,000-strong Allied Commando force killed or captured. After the withdrawal, the DB4s were abandoned on the beach.

Further small batches of the crawler were made up to 1945. Most were fitted with LaPlante-Choate bulldozers, although some may have had Bray equipment. Several of the crawlers were assigned to Operation Overlord - the Allied invasion of occupied France. After D-Day (6 June 1944), the DB4s were used to establish defensive positions for the breakout from Normandy. The crawlers featured in some of David Brown's advertisements and allowed the company to cri[b] from Churchill with the slogan 'Tractor versatility on the beaches, on the airfields, on the land an[d] in industry'.

Although the official production figure for th[e] DB4 was 110 machines, there remains some doub[t] as to the accuracy of this amount with rumour[s]

RIGHT:
This tractor, No.63, is one of several DB4 crawlers that David Brown rebuilt for agricultural use between 1946 and 1947. These machines are identified by their steel radiator grilles.

LEFT:
Registered in 1947, this DB4 is one of the crawlers that David Brown advertised as 'rebuilt for peace'. It is seen baling in Cheshire.

that up to 300 Dorman engines were delivered to Meltham. As only about a dozen tractors have survived into preservation it seems most likely that the factory figure is correct. Tractor serial number 110 is in the hands of a Devon collector, but there are no records of any later machines in existence. As a point of interest and to put the DB4 production into perspective, the machines built at Meltham were only a drop in the ocean when compared to the number of Caterpillar D4s made at Peoria, which was in the region of some 85,000.

During 1946 and 1947, David Brown repurchased a number of surplus military DB4

LEFT:
A DB4 crawler comes to the rescue during the severe winter of 1947. Two teams of tractors cleared a way through to Castle Hill to rescue catering manager Jack Holland's stranded family, recovering three Huddersfield Corporation buses at the same time.

ABOVE:
Jack Holland is reunited with his family on Wednesday 5 February 1947 after the crawlers succeeded in clearing a track to his house.

crawlers from the government through the Ministry of Agriculture. These tractors were returned to Meltham for a complete overhaul and may have accounted for some of the extra Dorman engines supplied. The reconditioned machines had their cast-iron grilles replaced with steel radiator grilles before most were resold to farms under ministry permits.

While at Meltham, some of these crawlers were called on to provide rescue work during the severe winter of 1947. On the evening of Sunday 2 February, David Brown's catering controller, Jack Holland, was returning to his home after supervising the arrangements for the night-shift canteen at Meltham. Very heavy snow had fallen, completely cutting off all the roads to his house at Castle Hill, an isolated site of ancient fortifications overlooking Huddersfield, and he had to spend the night in a hotel. Further severe drifting prevented him from getting home on Monday or Tuesday, and by Wednesday morning, he was becoming very concerned that his wife and children were running out of food.

Mr David Brown, hearing of his catering manager's plight, immediately sent out two teams of DB4 crawlers, some fitted with bulldozers, to clear a way through to Holland's house. The two teams worked in a pincer movement from different directions. By late afternoon, the crawlers had succeeded in clearing a track to the house, cutting through four miles of drifts and recovering three Huddersfield Corporation buses into the bargain.

In 1950, David Brown Tractors Ltd sold its stock of surplus DB4 parts, with the exception of the final-drive gears, pinions and flanges, to Tractor Spares Ltd for £2,000. Charles Weight believes that the undercarriage spares alone netted his company nearly $50,000 in sales over the next few years, which suggests that a good few DB4 crawlers were still operating somewhere in the world.

ABOVE:
One of only a handful of DB4 crawlers in preservation, this ex-military tractor still has the original cast-iron grille and the frames for the bulldozing equipment.

CHAPTER 5 — *Power for Peace*

David Brown's chief demonstrator, Nigel Palmer (centre), shows the prototype Cropmaster to an official from the War Agricultural Executive Committee (right) just prior to the new tractor's launch in April 1947.

Meltham's wartime contracts undoubtedly allowed the tractor business to grow. Whether it would have survived without them or just fizzled out to become another also-ran of the tractor industry, we can only speculate. The reality is that David Brown Tractors Ltd emerged from the conflict as a force to be reckoned with and during the post-war period it went from strength to strength.

If nothing else, the war gave the company the opportunity to gain valuable experience in the production of a wide range of models. Like the young lads who went to war, the tractor firm had an abbreviated adolescence and had to grow up very quickly. Within just six years of its inception, it had built everything from a basic agricultural tractor to an aircraft-towing vehicle and a heavy crawler. Ford had built far more tractors in the same period, but they were all based on the same tried and tested design that had been in production since 1917 with only minor variations. The same could not be said for Meltham's products, which had propelled David Brown into the big league of tractor manufacturers in a relatively short time.

The parent organisation, David Brown & Sons, also came out of the war in a strong position with its coffers boosted by the extensive defence contracts. The group, reinforced by the acquisition of the Muir Machine Tool Company of Manchester in 1944, became a founder member of NAVGRA, the Navy and Vickers Gear Research Association, two years later. Allan Avison was also actively exploring the possibilities of setting up subsidiary David Brown operations overseas, particularly in South Africa.

As the pressures of war production fell away, Mr David Brown became like a man released - freed from the constraints of military contracts and government intervention. Once the war ended, Lord Brabazon handed the chairmanship of the organisation back to David, who was free once again to run his own companies as he wanted to run them. Brabazon had become a close friend and agreed to stay on board as a director of David Brown & Sons.

Brown's uncanny knack of mixing business with pleasure without neglecting either was the envy of many. He obtained his pilot's licence in 1946 and occasionally flew the company's private plane, a seven-seater de Havilland Dove. A runway was laid out on Crosland Moor with a hanger erected next to the old implement shed. The same year saw David elected joint master of the Badsworth hunt; his passion for horses extended to a string of hunters and steeplechasers, which he both bred and trained. He always maintained that it was his ability to choose the right people to be his company lieutenants that allowed him time for all his interests. His simple philosophy was 'Get a good man and let him get on with it. If he is no good, get another man. But if he is a good man, support him.'

James Whitehead was one of Brown's 'good' men. The pair first met in the mid-1930s when Whitehead was employed by British Bedaux, a firm of consultants advising on the integration and reorganisation of several subsidiaries of the David Brown group. Brown was impressed with Whitehead's capable manner and asked him to act as manager of the Penistone foundry until a permanent appointment could be made.

During the war, Whitehead became David's personal assistant, taking on the responsibility for running the Meltham plant as acting general

ABOVE:
An ex-RAF VIG1/100 tractor is seen preparing the site for the runway for the company's private plane on Crosland Moor. Mr David Brown obtained his pilot's licence in 1946.

manager of David Brown Tractors Ltd after H. P. Scott moved on in 1942. In 1945, he started reorganising the plant, bringing it back from its wartime footing into the peacetime production of agricultural tractors and implements. He was assisted in his undertaking by Fred Marsh, the commercial manager. Their main priorities were the twin objectives of increasing both sales and production volumes, the latter standing at around twenty tractors a week in 1945.

The two managers laid down a production schedule for David Brown Tractors Ltd, and the company was praised by the then Minister of Agriculture, Tom Williams, during a visit to the plant on 2 November 1946, for being one of the few firms to do so. Williams was also impressed by the fact that, by the time of his visit, the company had already exceeded its planned production of 2,000 tractors for that year.

Marsh and Whitehead also initiated a survey that was carried out in the name of David Brown Tractors Ltd to determine an estimate of the post-war national demand for agricultural tractors. The estimate took into account many variables, including the number and size of agricultural holdings in the country, the rate of decline in the use of horses, the increase in demand for power on the land and the average working life of a tractor (which was estimated to be 7,000 hours or seven years).

It was an important and well-researched survey, which reached the conclusion that 40,000 new tractors a year would be needed by mid 1947. The estimate actually sidelined a similar survey carried out by the National Farmers Union, which

ABOVE:
The men at the top of David Brown Tractors Ltd in the immediate post-war period: James Whitehead (left), Fred Marsh and Mr David Brown, enjoy a working lunch in the staff canteen.

LEFT:
The engine assembly area in the tractor shop at Meltham. During 1946, David Brown built over 2,000 tractors and exceeded its own production schedules.

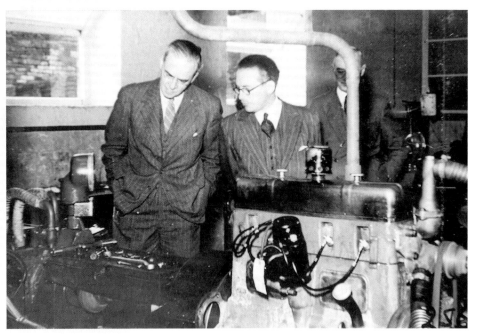

ABOVE:
Mr David Brown (right) shows Sir Stafford Cripps, the president of the Board of Trade, around the experimental department at Meltham. The company's enlightened approach to tractor production attracted a number of government officials to the works in the immediate post-war years.

RIGHT:
Farming author, A. G. Street (centre), is shown around the parts department during a visit to Meltham in 1946. Street, one of several leading agriculturists who expressed an interest in the company's activities, was a keen user of David Brown equipment on his own farm.

had put the demand for tractors at a lesser figure. David Brown's conclusions became the basis of the government's policy planning and actually resulted in an increase in the allocation of raw materials for agricultural machinery production. In the event, the average yearly output of tractors in Britain over the 1946/1947 period was just over 43,000 units, so David Brown's estimate was almost spot on.

The company's forward-thinking approach to tractor production saw Meltham regularly entertaining government officials such as Minister of Agriculture, Tom Williams, and the president of the Board of Trade, Sir Stafford Cripps. Several leading agriculturists and prominent machinery correspondents including A.G. Street, D. N. McHardy and T Hammond Craddock, also made repeated visits to the plant to see its production methods first-hand.

Although like chalk and cheese, Marsh and Whitehead made a very good team and were the driving force behind David Brown Tractors Ltd in the post-war years. Marsh was outgoing and enthusiastic, but his eagerness was tempered by Whitehead's more reflective and circumspect nature. Both were appointed to the board of David Brown Tractors in March 1946.

Other post-war appointments at Meltham saw Arthur Blackwell become works manager with Ronnie Marshall as assistant works superintendent. Following the departure of Captain Arlborough-Smith, Charles Birney took over as sales manager and was joined by Bill Barton as assistant sales manager and Tom Lazenby as sales assistant. Len Craven became parts department superintendent and reported to Birney. Vincent Gallagher was supply manager and

LEFT:
The leading figures in Meltham's engineering department attend a demonstration of the VAK1 tractor at Evesham in June 1944. The technical director, Dr Henry Merritt, is on the extreme left of the photograph while the chief designer, Herbert Ashfield, is on the extreme right standing next to the chief experimental engineer, Charles Hull.

had the unenviable task of procuring scarce raw materials in the face of tough market controls.

The engineering department at Meltham was still ruled by Henry Merritt at the end of the war, but the technical director seemed increasingly at odds with his managing director on matters of design and the two clashed regularly. At times, Merritt's work often seemed to be at a tangent to company policy and, as his relationship with Mr David Brown deteriorated, it looked increasingly as if his days at Meltham were numbered.

With Merritt concentrating nearly all of his efforts on the development of a completely new model, the improvements to the existing tractors and implements fell at Herbert Ashfield's door, the latter having been appointed to the position of chief designer in 1944. The work was in safe hands as Ashfield was proving to be a very gifted engineer and was successfully guiding through a series of updates to the VAK1 tractor.

The VAK1/A Tractor

The farm tractor works in a challenging and changing environment and for that reason its design will always be an inexact science. Its development and evolution is forever ongoing.

LEFT:
A VAK1 tractor that was used as a test-bed for several of the proposed VAK1/A improvements, including the new manifold, vaporiser and variable-speed governor. Note the bypass sleeve on the air-inlet pipe.

ABOVE:
The prototype vaporiser for the VAK1/A tractor showing the new manifold with the pre-heater. Note the cowling covering the inlet manifold.

RIGHT:
A prototype VAK1/A tractor on field test. The new model also incorporated changes to the front-axle layout to give the tractor a longer wheelbase to improve its weight distribution when handling heavy implements.

Once a model is released, the engineers will continually monitor it in the field to ascertain its strengths and weaknesses, gathering feedback information for improvements and to lay the foundation for future models.

Such it was with the David Brown VAK1, and areas for improvement and enhancement that had been identified as early as 1941 were incorporated into a series of planned changes to be introduced ahead of any future new model releases. The problem was that the defence contracts took precedence and the programme of improvements was delayed with many of the revisions eventually being modified or shelved. Wartime staffing changes and the coming and going of different consultants meant that nothing was ever finalised until Herbert Ashfield took over as chief designer and managed to draw all the strands together into one coherent design policy. His decisions were always based on sound engineering sense rather than fanciful ideas and the practicalities of the job were more important to him than deference to his superiors. By his own admission, he lacked experience with tractors in the early days, having come from a heavy-engineering background, but made up for this by accompanying Fred Meadowcroft on field tests and trials whenever possible.

The changes identified for the VAK1 included an improved vaporiser and governor, a greater choice of transmission speeds, stronger final-drives, an integral rather than an 'add-on' hydraulic unit and ultimately a diesel engine. The eventual plan had been to introduce most of these features on a completely new model, codenamed VAK2. This was the design that Merritt had been working on and

was the proposed replacement for the existing tractor. However, progress was slow and it was looking increasingly unlikely that this machine would ever go into production. As an interim measure, it was decided to upgrade the current VAK1 design, concentrating mainly on revisions to the manifold, air-cleaner, carburettor and governor assemblies.

The new one-piece combined inlet and exhaust manifold was the long-awaited improved vaporiser that was the culmination of work begun by Albert Kersey and carried on George Roesch. It was designed to give a better mixture, increased power and a more rapid changeover to TVO. Air entering the oil-bath cleaner was drawn from a pre-heater on top of the exhaust manifold. This warmed the air before it was cleaned and mixed with the fuel in the carburettor. The mixture was then completely vaporised by an automatic load-controlled 'hot-spot' in the manifold.

The automatic 'hot-spot' was designed to optimise the exhaust circulation through the manifold and was controlled by a flap with a bob-weight. Charles Hull fitted the prototype manifold to a tractor in the experimental workshop. The engine was then run under load to determine the design and stress of the flap for optimum performance before the findings were passed over to the engineering department. The manifold design was a simple but very efficient arrangement.

The cowling on the inlet manifold was also designed to guard against draughts from either the cooling fan or side winds. A sleeve around the air-inlet pipe could be rotated to uncover an aperture in order to regulate or bypass the pre-heater depending on weather conditions or the working temperature of the tractor.

The earlier two-speed governor, controlled by a rod, had operated in conjunction with a throttle lever on a ratchet. It was a cumbersome and restrictive arrangement and the driver had to adjust two levers to set the engine speed. The new all-speed governor was operated by a single throttle lever. The governor was more precise and the engine speed was proportional to the movement of the lever, which allowed a constantly-variable adjustment between the range of 600 to 2,400 rpm. The throttle lever had a catch-stop at the normal working speed of 1,300 rpm.

Henry Merritt supervised the changes to the engine governor, but left it to Herbert Ashfield and others to pull all the modifications and refinements together into one revised model and ensure that it all worked in the field. This greatly improved tractor was to be known simply as VAK1/A.

LEFT:
A prototype VAK1/A tractor outside the experimental workshop at Meltham. Improvements had been made to the vaporiser, carburettor, front axle and final-drive gears.

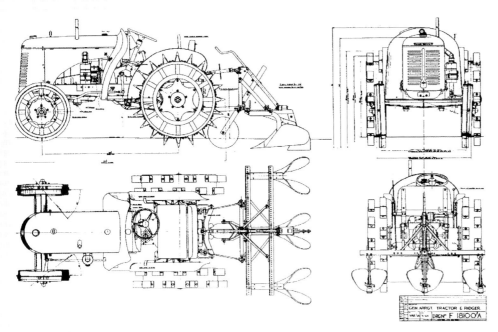

ABOVE:
A drawing showing the general arrangement of the VAK1/A tractor and ridger. The VAK1/A's power lift was superseded by the U36A power unit soon after the tractor's launch.

Pre-production field trials with the VAK1/A in early 1945 highlighted problems with the new governor in that it had the wrong weights, which meant it was sensitive but under-powered. It struggled against the friction of the throttle spindle attached to the butterfly in the carburettor. This caused hunting - irregular running of the engine - particularly at idling speed with the governor pulling the throttle shut. Herbert Ashfield took the carburettor to the experimental department to discuss the matter with Charles Hull.

Hull stripped down the carburettor and machined a groove into the throttle spindle to take needle roller-bearings, which he fitted together with felt seals to keep the dirt and air out and prevent surging. The spindle now moved freely, the engine ran evenly and was more responsive to the throttle. From then onwards, every Solex carburettor fitted to a David Brown tractor was modified in the same manner. A cam was also fitted to the governor control-rod to act as a damper for the governor and even out the idling speed.

The carburation of the VAK1/A was also altered to make the engine easier to start. The new manifold arrangement meant that the carburettor was now 3 in. lower. This allowed a sufficient head of fuel to be maintained by gravity and ensured that there were no fuel starvation problems when working on gradients. The need for a fuel pump was eliminated, which the company claimed would save valuable petrol during the changeover period.

The VAK1/A engine was essentially the same as the VAK1 power unit except for an improved and simplified engine lubrication system. This dispensed with much of the internal pipework and the relief valve was now integral with the oil pump. The VAK1's two drain-taps for the cooling system, positioned to the front and rear on the nearside of the block, were also replaced by a single centrally-mounted drain screw behind the carburettor.

The durability of the transmission was improved by changing the tooth shape of the final-drive gears. The ratio of these gears was also altered to give approximately a 5 percent reduction in speed to increase the tractor's drawbar pull and reduce the loading on the transmission. This was done to meet the requirements of the Ministry of Agriculture which had requested that all medium-weight tractors to be built in the immediate post-war period should be capable of handling a three-furrow plough.

Visually, the VAK1/A was almost identical to the VAK1. The pre-cleaner dome had disappeared following the changes to the manifold and the air-intake system, and a new steel front grille with horizontal slats had superseded both the earlier cast-iron and 'bullet-hole' grilles, but otherwise the new model looked very much like its predecessor. What was not immediately obvious was a number of changes to the front-axle arrangement.

The front axle had been turned round with its trunnion mountings reversed. This had the effect of moving the axle forward and giving the VAK1/A a longer wheelbase by some 4½ in. (an increase from 64½ to 69 in.). The axle was also strengthened. Out went the old longitudinal tube and in came a new box-section form fabricated from steel plate.

The longer wheelbase improved weight distribution (so that the tractor could better handle a three-furrow plough) and the forward position of the front wheels protected the radiator grille in case of collision. This had been a particular problem with the early VAK1 model.

LEFT:
A pre-production VAK1/A tractor with oversize rear tyres. Note how reversing the position of the front axle has moved the wheels forward to protect the radiator and grille in case of a collision.

with the cast-iron grilles, which had been easily broken. With the axle set back and the nose of the tractor protruding beyond the wheels, the grille had been vulnerable to damage, which is why few of the VAK1s in preservation today have survived with their original radiator grilles intact.

The VAK1/A's serial numbers followed on in sequence to the VAK1 series. The first production VAK1/A, serial number 6351, was made on 6 March 1945 and the model was unveiled to the press and public the following month at around the time of VE (Victory in Europe) day, making it David

LEFT:
A 1945 David Brown VAK1/A tractor. The first of the improved models was made in the March of that year. Initially, the tractor was only available on steel wheels, but most owners converted their machines to pneumatic tyres once the restrictions on rubber ended.

RIGHT:
The David Brown VAK1/A tractor and its associated equipment is launched in Ireland at the Dublin Spring Show in May 1945.

Brown's first post-war tractor. The company was at great pains to point out that the improvements incorporated into the new model came at no extra cost. There had been no price increase and the new tractor cost no more than the superseded VAK1 at £351 complete with the U36 power lift. The basic model without hydraulics was priced at £300.

Continuing post-war shortages meant that the VAK1/A was initially only offered on steel wheels. By November, the restrictions on rubber had eased and the David Brown tractor was once again available on pneumatic tyres for an extra £28 over the basic list price.

Soon after the VAK1/A was launched, the U36 power lift was superseded by the U36A hydraulic

RIGHT:
The David Brown VAK1/A was available with pneumatic tyres from November 1945. The tyres cost an extra £28, taking the list price of the agricultural model with power lift to £379.

LEFT:
Herbert Ashfield field-tests the VAK1/A tractor, ploughing in the rain at Bolton Abbey in December 1945. This farm in Wharfedale, owned by William Foster, was regularly used by David Brown for field trials.

unit with extended lower-hitch brackets and an adjustable top-link. These minor changes were made to improve the performance of the ridger and cultivators when used with the single depth-wheel.

Problems with the rotary valve in the power lift led to Fred Meadowcroft fabricating a spring-loaded ball-valve, which was more foolproof and simplified the hydraulic control-valve mechanism. It also eliminated the earlier problems with the hydraulic oil heating up because the design of the new valve meant that it was no longer under continuous pressure. The valve was patented (under both the company's and Meadowcroft's names) and fitted to all power lifts built to U36B specification from 16 September 1946. The control lever for the U36B lift was bolted to the offside mudguard rather than to the unit itself.

During 1946, the company changed its policy of fitting coil and distributor ignition to tractors supplied with electric starting and lighting. From 4 June, all tractors were equipped with magnetos (the Lucas 4VRA was now standard), although coil ignition was still offered as a special order under reference U47.

Other accessories (or 'auxiliaries' as David Brown liked to call them) introduced with the VAK1/A included the U24 steel wheels with retractable strakes and the U27 ridge-riding wheel. The latter was a special steel front wheel, launched in March 1945, with a vee-shaped rim for running on top of the ridge when splitting back with the ridger.

Owners of the VAK1 tractor could take advantage of the VAK1/A improvements as most of the modifications could be retro-fitted to the earlier model. An engine conversion exchange scheme was also introduced, offering a complete service-reconditioned VAK1/A engine with the new manifold, air-cleaner, carburettor, governor and control lever plus a kit of all the necessary parts to effect the changeover to the improved specification. Many VAK1 owners also replaced their cast-iron (and often broken) radiator grilles with the new steel VAK1/A grilles to give their

RIGHT:
This VAK1/A formed part of an order received in May 1945 from the Norwegian government for tractors to operate in the Finnmark region inside the Arctic Circle. The only modifications made to the tractors to cope with the arctic conditions were the fitting of a heavy-duty electrical system and wider wheels with oversize tyres.

tractors a more modern appearance.

One interesting variation on the VAK1/A was a narrow version that was manufactured by Stanhay Ltd of Ashford in Kent under the VAK1/A/H designation. It was launched in November 1945 and was aimed at the hop and fruit growers in Herefordshire, Worcestershire and Kent.

The narrow model had its front axle turned around and relocated in the original VAK1 mounting position to reduce the turning circle for use in hop alleys. The axle was also cut and shortened so that the tractor, if used with rowcrop steels on the rear, had an overall width of only 4 ft 6 in. Steel cone-covers were fitted to the front wheels to avoid damage to hop bines or fruit bushes. The tractor could be matched to Stanhay's Paraframe underslung hoe and rear-mounted toolbar with hop-garden ploughs.

The VAK1/A had the distinction of being the first David Brown tractor to be sold overseas after an order was received from Norway in May 1945, just after the new model had been launched. The tractors were required for Finnmark, the most northerly area of the country and some 200 miles inside the Arctic Circle. The Finnmark region had been burned and destroyed by German army units retreating from Russia and Finland, leaving an area of scorched earth larger even than the whole of Denmark.

After Norway was liberated, its coalition government put plans in place to rehabilitate Finnmark and get as much land under the plough as possible to support the indigenous 'Sami' or Laplander communities which produced milk, meat and potatoes. Representatives of the government approached the British authorities with a view to obtaining a consignment of suitable tractors for working on the difficult terrain. Much of Finnmark was on a plateau 1,000 ft above sea level where the temperatures regularly dropped to 35 degrees Fahrenheit below zero. The tractor would have to clear 10 ft high snowdrifts as well as cultivate land.

After due consideration and exhaustive investigations, the Norwegian authorities decided on the VAK1/A, partly because of its ease of servicing. The wags at the Meltham plant liked to suggest that the David Brown tractor was chosen because Finnmark's climate was not much different from that experienced on the moors in Yorkshire! There was probably a grain of truth in this because no modification had to be made to the tractors to cope with the arctic conditions other than fitting a heavy-duty electrical system, wider wheels and oversize tyres.

The Finnmark order was completed in only seventeen days. The tractors, implements and spares were packed into heavy wooden packing cases for shipment. To overcome language difficulties and to make allowances for inexperienced operators, coloured labels were attached to each filling aperture on the tractors to correspond with similarly-labelled containers of lubricants. For months afterwards, an unsubstantiated rumour circulated the works that the first shipment to Norway had been unfortunately torpedoed by an isolated U-boat, unaware that the war had ended.

The popularity of the VAK1/A, particularly after it was released on pneumatic tyres, was such that demand continually exceeded supply. In November 1945, David Brown Tractors Ltd notified its dealers that, despite increased production, it could no longer guarantee delivery in less than twelve months from the date of receipt of the necessary permit from the War Agricultural Executive Committee, whose approval was still needed to secure a new tractor. The company suggested that the dealers tell their customers that 'the David Brown tractor is worth waiting for'.

A marked expansion in productive capacity saw output jump to thirty-six tractors per week in 1946, but it was still not enough to satisfy orders. The VAK1/A was only really a stopgap machine, but it remained in production for a lot longer than expected until finally being replaced by the VAK1/C in mid-1947. A total of 3,502 VAK1/A tractors were built, and the last one, serial number 9852, rolled off the line on 27 June 1947.

ABOVE:
The Finnmark order was completed in seventeen days and tractors, implements and spares are seen crated at Meltham ready for shipment to Oslo.

LEFT:
Part of the Finnmark consignment of tractors leaving Meltham. The shipment represented David Brown's first overseas order. The Atkinson lorry belonged to a local haulage contractor.

RIGHT:
VAK1/A models on the assembly line at Meltham. By 1946, production had risen to around thirty-six machines per week but there was still nearly a twelve-month waiting list for David Brown tractors.

The David Brown implements were similarly affected by supply difficulties as allocations of raw materials upset production schedules. By the beginning of 1946, certain types of plough were available for immediate delivery and most of the outstanding orders for cultivators had been cleared. However, almost the entire coming year's output of potato spinners had already been taken up.

Only three machines were added to the implement range during the VAK1/A period. The first of these was the MR1 cutterbar mower. Like the potato spinner, the origins of this machine could be traced back to the end of the Ferguson-Brown era, but its development had been delayed by the war. Designed by Fred Meadowcroft, it was eventually launched in May 1946 following extensive trials at Eastfield Farm at Haxby near York. Farmed by the Agar Brothers, this was one of two sites in Yorkshire regularly used by Meadowcroft for field tests of the tractors and implements. The other was a farm owned by William Foster at Bolton Abbey in Wharfedale.

The mower was mounted on

RIGHT:
Designed by Fred Meadowcroft, the David Brown MR1 cutterbar mower was launched in May 1946. It was attached to the tractor via a sub-frame and the wire rope and pulley arrangement for lifting the cutterbar can just be seen behind the tractor's rear wheel.

ABOVE:
The David Brown SU1 sprayer is introduced at a demonstration in Lincolnshire staged by the local distributors, Belton Bros & Drury, in early 1946. The sprayer was manufactured by Patrick & Wilkinson of Belfast.

to the tractor via a sub-frame. It was somewhat cumbersome and fitting took twenty to thirty minutes. The 5 ft cutterbar was driven by vee-belts from the power take-off shaft and raised by a wire-rope attached to the tractor's lift arms. The machine was not entirely successful during its first season and several modifications were made for 1947, including new fingers, shoes, and swath board as well as the provision for increasing the tension of the vee-belts. This improved model cost £72.

The David Brown SU1 'Unit Principle' sprayer was introduced in April 1946, following demonstrations in Lincolnshire, Lancashire and Northern Ireland, but deliveries did not commence until the following June. Unlike the other implements in the range, the sprayer was not made at Meltham but was designed and manufactured for the company by Patrick & Wilkinson of Belfast.

Patrick & Wilkinson made a number of attachments for David Brown tractors and the SU1 was based on its 'Sprayquick' Universal Sprayer. It had an 80 gallon tank, seven lances and would cover a 16 ft bout. The sprayer took 30 minutes to fit to the tractor and could spray 30 acres in an eight-hour day. It was priced at £95.

The third of the implement introductions was

LEFT:
The David Brown BR1 beet lifter was designed in conjunction with Belton Bros & Drury and is seen on trial in Lincolnshire. The two-row lifter was launched in October 1946.

the David Brown BR1 beet lifter, which was designed in conjunction with Belton Bros & Drury for farmers in the eastern counties. It was launched at a demonstration of harvesting machinery staged by the Sugar Beet Education & Research Committee on 24 and 25 October 1946 at Bury St. Edmunds in Suffolk. It was a two-row lifter with two pairs of lifting shares and disc-coulters attached to a heavy toolbar frame.

VAK2 and the Nuffield

As mentioned previously, Henry Merritt spent the last years of the war developing a completely new tractor that was eventually expected to replace the VAK1. His work was kept separate from and ran parallel to the interim improvements to the existing agricultural model that resulted in VAK1/A.

The tractor, provisionally designated VAK2, only took shape very slowly, which is why Mr David Brown, frustrated by his technical director's lack of progress, often brought in outside consultants to assist. At one time, a team of Vauxhall engineers were invited to Meltham, but all they wanted to do was write off the whole David Brown tractor development programme so they were soon sent packing.

As the VAK2 project progressed, Merritt produced copious plans of different arrangements for the new machine and was assisted in the drawing office by Steve Moorhouse. The tractor was developed in secrecy with most of the prototype parts handmade in the experimental workshop and then hidden away in the storeroom.

Because of the secrecy surrounding the project, little is known about the tractor and the few details that have emerged are based on workshop gossip. However, it seems that the idea behind VAK2 was to produce a larger and more powerful machine (around 10 hp more than the VAK1). It would have a greater number of gears, an integral hydraulic lift and independent power take-off. The possibility of using a stressed engine block like that of the Fordson to join with the transmission to form the backbone of the tractor, thus making Kersey's concept of the cast-iron mainframe redundant, was also considered. Another idea was to use a half-mainframe at the front of the tractor to support the engine. No doubt, a diesel version was also high up on the list of priorities.

Only one prototype was ever made. Assembled by Charles Hull and his team in the experimental workshop in early 1945, it was something of a transitional machine. It was built to test out some

RIGHT: *This is the sole prototype of the David Brown VAK2 tractor, a project conceived by Dr Henry Merritt. Its features included a six-speed gearbox, a new front-axle arrangement and a centrally-mounted steering column going through the middle of the fuel tank. Note the air-cleaner body incorporated into the front-half mainframe.*

of the new features, including a six-speed gearbox, but still incorporated a number of VAK1 characteristics.

The prototype had a completely new front axle, mounted on a trunnion bracket and bolted to the underside of the frame in two positions, which was designed to make the tractor more easily convertible for rowcrop work. It also had a different steering arrangement using a new design of worm-and-nut steering box that was centrally mounted with the steering column going through the middle of the fuel tank.

The tractor had an integral power lift with a new design of rotary gear-type pump. The rotors of the pump ran on a special type of roller-bearings mounted within the pump housing. The delivery gallery and ports were integral with the pump housing, which also contained a combined control and pressure-relief valve.

The prototype was powered by a VAK1 engine but was fitted with a new front-half mainframe that incorporated the air-cleaner body, which was built in below the carburettor, thus giving the tractor an uncluttered appearance. None of the field-test engineers can remember the machine ever being used for trials and it is believed that it was returned to the experimental department and partly dismantled.

By this time, it was becoming evident that Merritt's designs were too radically different from the existing tractor layout to be of much use to the company at this time. His proposals for VAK2 included a more powerful 40 hp engine and a new design of gearbox frame that could not be altered to fit the existing machine tools. Tooling up for such a completely new design was going to be prohibitively expensive and would also cause disruption to the assembly line and affect output.

The proposed design brought the underlying tension between Mr David Brown and his technical director to a head. The VAK2 project was dropped and Merritt left the company to join the Nuffield Organisation in August 1945. Fred Hollister, who had earlier been recruited by Merritt from Armstrong-Siddeley, then took over as Meltham's technical manager. VAK2 was left half-built, sitting on wooden blocks in the corner of the experimental department until it was finally scrapped in 1946. That could have been the end of the story for this particular design were it not for the fact that Merritt decided to draw on some of the ideas he had for VAK 2 when he went to Nuffield.

The Nuffield Organisation, founded by Viscount Nuffield (formerly William Morris), had a number of leading British automotive concerns, including Morris Motors, the MG Car Company, Riley Motors, Wolseley Motors and SU Carburettors, under its parentage. It was no secret that the organisation had been considering tractor production for some time and Merritt was the ideal recruit to head the design team.

Merritt's relationship with the organisation had begun during the war while its Nuffield Mechanisations subsidiary had been producing his tank gearbox. He had remained in contact with Nuffield's vice-chairman, Sir Miles Thomas, and had even helped sort out a problem with suspension on a Morris car. Thomas later revealed that the Nuffield tractor project had begun in early 1945 and there is little doubt that Henry Merritt was approached while still working at Meltham. Cynics have suggested that he had secretly agreed a deal with Sir Miles and had been using David Brown's resources to develop his designs for the new tractor before joining Nuffield, but most of the engineers who worked with him at Meltham deny that this was the case.

Henry Merritt was made senior technical executive of the Nuffield Organisation, joining author and agriculturist, Claude Culpin, who had initiated the tractor project and had carried out research into what farmers required from a new machine. The design team was installed in an office just outside the car finishing shop at the Wolseley Works at Ward End in Birmingham and work on the first prototype began on 6 September.

One of Merritt's first concerns was finding equipment to be used with the tractors and establishing a line of implements. Having little knowledge of what was available on the agricultural machinery market, he decided to start with his old firm. On 30 January 1946, he wrote to David Brown Tractors Ltd asking about the possibility of procuring implements. A similar request was made to Ransomes, Sims & Jefferies.

The answer Merritt got from Meltham was probably not what he expected. Instead he received a rather terse reply, ignoring his request and expressing 'broader concerns over matters'. The broader concerns were not just over Merritt's involvement with the Nuffield project, but also his alleged poaching of personnel from David Brown.

On 18 March, Sir Miles drafted a letter to David Brown denying that his company was trying to poach staff. It was not exactly true; Nuffield, like many other engineering concerns in the post-war period, was critically short of designers. To make up for the deficit, Merritt had persuaded three or four of his former colleagues from Meltham, including draughtsmen, Steve Moorhouse and Jack Crichton, to join him on the tractor project. It was an easy decision for Moorhouse; he had always thought very highly of Henry Merritt and found Fred Hollister, his new boss at Meltham, a difficult man to work for.

During March, Sir Miles Thomas gave a press conference to announce details of the new Nuffield Universal tractor and released an artist's impression of the proposed machine. When the illustration appeared in the trade journals, David Brown was horrified to discover that the proposals for the new tractor bore more than a passing resemblance to the VAK2. The company's solicitor, Max Ramsden, immediately dispatched a letter to Thomas, suggesting that Merritt and the Nuffield Organisation had stolen its ideas.

This sparked off a series of communication between Max Ramsden and Nuffield's solicitors Hubert Gower, with the former arguing David Brown's concerns over the piracy of its design and the latter rebutting all allegations. Eventually Hubert Gower asked Merritt to consider the technical aspects of the dispute. He conceded that three features of the new Nuffield Universal notably the front-axle arrangement, the steering column through the fuel tank and the method of mounting the rotor for the hydraulic pump, were similar to the VAK2 design, but added that they were not covered by David Brown patents.

In July, Merritt, through Hubert Gower released a statement highlighting the technical points in which the Nuffield tractor differed from the David Brown design. These included its appearance, power-unit, number of gearbox speeds, altered wheel arrangement, ground clearance, steering brakes, hand clutch, power lift and overload release mechanism. Merritt also stated that he 'used his talents at David Brown to overcome deficiencies and defects in company design' - which was hardly likely to endear him to his old firm and was a little like adding salt to the wound.

Mr David Brown asked Herbert Ashfield to appraise the Nuffield design. Ashfield concluded that it was a medium weight machine, aimed more at Ford's market than David Brown's, and assured his boss that there was nothing to fear. Finally, after months of lengthy correspondence, Max Ramsden let the dispute with Nuffield quietly drop.

Meanwhile, Nuffield had its first prototype ready for trials by June 1946 and Merritt had

BELOW:
Released in March 1946, this artist's impression of the proposed Nuffield tractor bore a striking resemblance to the VAK2 prototype and sparked off a war of words between David Brown and the Nuffield Organisation.

LEFT:
Fred Meadowcroft at the wheel of one of the prototype Nuffield tractors at Four Oaks near Birmingham in 1946. A three-furrow David Brown plough was used for many of the field trials.

managed to persuade Fred Meadowcroft to join him and organise the field-test department. Hollister had tried to keep Meadowcroft at Feltham but had refused to match the salary increase that the Nuffield Organisation was offering. As many as twelve prototypes were released for field testing during 1946 and several implements were used during the trials, but most of the ploughing was done with a three-furrow David Brown plough.

A report prepared by Merritt in October 1946 illustrates the high esteem in which the former David Brown men were held at Nuffield. Fred Meadowcroft is referred to as 'a very expert tractor and implement user and critic with constructive ideas' while Steve Moorhouse is described as a 'good technician and draughtsman'.

Another development saw the Canadian Massey-Harris concern, eager to source tractors for its British subsidiary, offer to distribute the Nuffield Universal in the UK and certain overseas territories. The proposal was declined in April 1946, after which Massey-Harris made a similar approach to David Brown in February 1947.

The offer was that Massey-Harris would stop selling its tractors in the UK and Ireland in exchange for the exclusive distribution rights to David Brown tractors in those same areas as well as in other territories where trade embargoes on the sale of the Canadian-built machines existed. It was also proposed that Massey-Harris would sell both its own and the David Brown tractors side by side in most European countries.

Not unexpectedly, David Brown also rejected the offer. It had not yet opened up its export markets to any great extent, but it had plans in hand and already had a loyal and strong dealer network of its own in the UK. Massey-Harris was left with no option but to begin manufacturing its own tractors.

In early 1947, Nuffield's development programme was temporarily abandoned, partly because of material shortages and partly because problems in sourcing castings from outside suppliers were making it difficult to produce the tractor at a competitive price. For the time being, Merritt and his team of engineers were seconded to the advanced car design section in the main Morris Motors drawing office at the Cowley plant near Oxford, where they mixed with the likes of Alec Issigonis. Steve Moorhouse, unhappy with the move and not wanting to work on car designs, left soon afterwards to join the new British

RIGHT:
An early production Nuffield tractor, an M4 model without hydraulic lift, outside the experimental workshop at Wolseley Motors. By the time the Nuffield was launched in November 1948, it bore few characteristics of the David Brown tractor.

International Harvester factory at Doncaster.

Fred Meadowcroft also refused to move over to the car side and decided to take a sabbatical, agreeing to return once tractor production was underway. In the meantime, he went back to Meltham and set up his own private enterprise, developing and manufacturing what was to become the world's first trailer tent.

Ironically, it was David Brown's detailed surv[ey] and estimate of the national demand f[or] agricultural tractors, which persuaded t[he] government to stimulate production and ea[se] steel supplies, that probably led to the Nuffie[ld] project being resurrected in September 1947. T[he] Universal tractor went back into production a[nd] was finally launched in November 1948. It now bo[re]

RIGHT:
A Nuffield Universal Four at work. Several former David Brown engineers, including Dr Merritt, Steve Moorhouse and Fred Meadowcroft, were involved in the Nuffield project.

little resemblance to a David Brown other than the front-axle arrangement, offset driving position and the use of a front mainframe extension to support the engine.

Meadowcroft returned in early 1949 to act as sales promotion engineer, handling field trials, demonstrations and even being sent to Canada to study methods of agriculture and oversee field tests in Saskatchewan. In about 1954, he resigned to join Fisher Humphries, one of Nuffield's approved implements suppliers, to develop a disc plough.

Henry Merritt left the Nuffield Organisation in 1949 and joined the British Transport Commission as chief research officer. Two years later, he was appointed chief administrative engineer to the Rootes Group at its Coventry motor factories, where he remained until he retired. He was still recognised as one of the world's leading exponents of gear design and was described as being 'in the front rank of authorities' in mechanical engineering.

Industrial Duties

The victory in Europe brought an end to the Air Ministry contracts for towing tractors, but the heavy industrial model did not disappear completely and a few trickled out of Meltham in the immediate post-war years. David Brown built over 400 industrial tractors during the VAK1/A period, but most were either IND series machines (without the turbo transmitter) for civilian use or were reconditioned Air Ministry machines.

Following the launch of the VAK1/A, David Brown issued a photograph of a 'medium industrial' version of the agricultural model. No specifications were ever released and all we can tell from the picture is that it had oversize tyres, heavy-duty electric starting and a lighting kit. It has been suggested that the tractor shown was in fact one of the Finnmark consignment and that the photograph was issued to test the market and see what interest it generated from highway contractors and municipal corporations requiring a light industrial machine. There is no evidence of any such version of the VAK1/A being put into production, probably because the company already had problems meeting the demand for agricultural models.

Of greater concern to David Brown was finding a peacetime role for its Air Ministry models. Towards the end of the war, the company developed and tested prototype versions of the VIG1/100 for logging, threshing and even general cultivations. The threshing version was very successful, having proved itself capable of hauling heavy threshers in tandem with balers in several hilly districts, and the decision was taken to include the model in the post-war David Brown range.

LEFT:
This photograph of a 'medium industrial' version of the VAK1/A tractor was issued in 1945, but the machine appears not to have gone beyond prototype stage. It had oversize tyres, electric starting and a lighting kit.

RIGHT:
A VIG1/100 heavy-industrial tractor being used for logging at Harewood House in Yorkshire in early 1947. After David Brown's Air Ministry contracts came to an end, the company tried to find a peacetime role for its industrial machines, organising a series of demonstrations to illustrate the tractor's versatility.

BELOW:
The David Brown Heavy Duty Tractor (Threshing Model) was launched in May 1946. Based on the wartime industrial tractor, it was aimed at contractors and was designated VTK1.

Aimed at farmers and contractors, the new machine was officially known as the David Brown Heavy Duty Tractor (Threshing Model). Launched in May 1946, it was designated VTK1 and priced at £750. Only limited numbers were available, and then only against a War Agricultural Executive Committee permit.

The threshing model was based on the VIG1/100 but featured a TVO engine with all the latest VAK1/A modifications. Its David Brown winch, sprag, four-point rear towing hitch and front-mounted shunting socket were all ideal for manoeuvring a threshing drum or recovering on soft ground. Heavy-duty electric starting (with two 6 volt CAV batteries) was standard, as was lighting, which the company claimed would provide adequate illumination for 'early-morning threshing and evening haulage'.

A belt-pulley with its own gearbox was mounted on the front nearside of the chassis frame. Its drive was taken by shaft from the front of the winch chain-housing and was engaged by a sliding-pinion dog-clutch operated by a lever on top of the pulley gearbox. Belt speed was 2,350 ft per minute at an engine speed of 1,600 rpm.

The VTK1 was only on the market for a couple of years and around 85 were built. The big drawback to sales was its price, which was more than double that of a standard agricultural tractor. Also, it was a heavy machine of limited use to the farmer other than for haulage or belt-pulley work, whereas a normal farm tractor could be used for a variety of operations.

The IND series industrial

LEFT:
The VTK1 threshing model (right) with its sister VAK1/A agricultural tractor. Priced at £750, the VTK1 was a very expensive machine and only 85 were built.

BELOW:
Herbert Ashfield and his colleagues inspect a David Brown threshing tractor at work. The VTK1 was fitted with a front-mounted belt-pulley, rear-mounted winch and electric starting and lighting. The pulley was driven by a shaft taken from the winch chain-housing.

RIGHT:
Nigel Palmer demonstrates the VTK1 threshing tractor to a group of farmers at Carmarthen in Wales in 1947. The control lever for the dog-clutch that engaged the belt-pulley can be seen on top of the pulley-gearbox housing.

BELOW:
David Brown's IND series of heavy-industrial tractors were supplied to a number of customers in the immediate post-war period, including Bertram Mills Circus.

tractors were also upgraded to incorporate the latest VAK1/A modifications, but were of course fitted with petrol engines. Customers for these heavy industrial machines included Rolls-Royce, Commer Cars, Brook Motors, Bertram Mills Circus, Derby Council and Glasgow Corporation. A small number of aircraft-towing tractors with the turbo transmitter transmission were also supplied to civil aerodromes and defenc[e] equipment manufacturers such as Bristol Aircraf[t] Handley Page and Vickers Armstrong. At leas[t] three of the towing tractors were shipped to th[e] Middle East during 1946 for use on airfields i[n] Kuwait.

Immediately after the war, David Brow[n] repurchased many of the earlier Air Ministr[y]

LEFT:
Derby Council bought this David Brown heavy-industrial tractor in 1947. Its duties included tar spraying and road maintenance.

tractors, both directly from the Ministry of Supply and through government disposal auctions. These machines were returned to Meltham to be reconditioned and brought up to the latest specifications before being resold either as threshing or industrial tractors and accounted for a good percentage of the company's industrial sales in the post-war years.

LEFT:
This 1946 David Brown heavy-industrial tractor was used by Glasgow Corporation's housing department to move prefabricated buildings around the city.

The David Brown Tractor Story

RIGHT:
A David Brown industrial tractor on trial with Huddersfield Corporation, recovering a broken-down trolley-bus.

Several David Brown industrial tractors were called on to rescue stranded vehicles on the moors around Meltham during the severe winter of 1947, which brought mixed blessings to the company. The so-called 'big freeze' or 'great frost' began on 24 January with heavy snowfall. It then snowed every day until 16 March. The snow was accompanied by sub-zero temperatures, and drifts of up to 10 ft deep blocked the roads. Coal could not get through to the power stations leading to power cuts; industry was paralysed and factories and steel mills were forced into short-time working.

Like all other manufacturers, David Brown was affected by the consequent shortage of fuel and raw materials. Meltham cut its working hours and actually had to suspend some of its clerical staff due to the subsequent decline in production. After the plant was forced to shut for a day on 10 March the rumour quickly circulated that David Brown Tractors Ltd was closing down for good!

The company was quick to deny the rumour and issued the following statement: 'While it has been extremely difficult to carry on owing to shortage of fuel and of materials, production has never stopped at Meltham and, in fact, we are probably the only major tractor firm in the country to continue during the crisis.' In truth

RIGHT:
A consignment of David Brown machinery for the Middle East, including three industrial tractors destined for use on aerodromes in Kuwait, leaves Scarr Bottom by rail in 1946.

LEFT:
David Brown put this heavy-industrial tractor on the road in 1947 and used it for its own internal haulage, such as handling a fuel bowser. The tractor is seen at the Farsley works where most of the industrial machines were reconditioned.

the order books were full and once the crisis was over, production was stepped up to around fifty tractors a week.

The up-side of the big freeze was that it brought the company plenty of publicity, both locally and nationally, such as the rescue of Jack Holland's family with the DB4 crawlers as previously mentioned. Very often, David Brown had the only equipment suitable for coping with the conditions, and this did not go unnoticed by the Halifax Corporation Transport Department, who immediately ordered six reconditioned Air Ministry tractors for snow clearance and road gritting in order to keep the Pennine routes open. These machines were fitted with enclosed cabs and Bunce snow-ploughs.

LEFT:
David Brown supplied six reconditioned Air Ministry tractors to the Halifax Corporation Transport Department in 1947. They were fitted with Bunce snow-ploughs and used to keep the Pennine routes open through the winter.

RIGHT:
This strange contraption was developed at Farsley for the Halifax Corporation in 1947. Its exact purpose is not known.

BELOW:
The prototype integral lift, which was originally conceived under the VAK1/B designation. The lift assembly was contained in the rear-axle housing and the pump was mounted on the front end-plate of the gearbox. The field-test engineers preferred a pan seat to the bench seat because it gave them a better view of the implement in work.

Cropmaster Farming

It is hard to believe that by 1946 David Brown had only been making tractors for ten years, but in that time it had built over 10,000 machines. The company had matured into a major player on the UK market and had a wide range of experience in developing different models for different needs. The engineering department at Meltham, depleted by the loss of some key personnel to Nuffield, was working at full stretch on a variety of developments from diesel engines to crawlers and had new agricultural and industrial models in the pipeline.

After VAK2 was dropped, it was decided to follow the VAK1/A modifications with a series of phased upgrades to the existing tractor rather than introduce a completely new model that would disrupt production schedules. Plans were put in place for VAK1/B (an integral power lift) and VAK1/C (a six-speed gearbox) as well as other future improvements including a diesel engine. Because of the high level of demand for the improved VAK1/A model and the desire not to interrupt production, the VAK1/B upgrade was delayed and eventually shelved.

The new technical manager, Fred Hollister, was an ebullient character who did not always get on well with his subordinates. He has been described as a 'belt and braces' engineer who tended to take the cautious approach, insisting on his staff adding extra strength to the machines to the point where they often

LEFT:
Charles Hull at the wheel of a prototype David Brown tractor incorporating the integral lift in early 1946. This machine was used as the basis of the VAK1/C design and gives a good impression of what VAK1/B would have looked like had it gone into production.

became cumbersome and impractical. His approach did not sit too well with his two top men, Herbert Ashfield and Charles Hull, who had both been brought up in precision engineering and believed that simplicity was the key to a good design.

Uncomfortable with his new boss, Hull eventually resigned in May 1946 to take up a new position with Simms Motor Units of London. The job had been offered to him through his contact with Professor Wisner and involved working on fuel injection equipment. Ashfield, on the other hand, managed to cope with the situation by circumventing Hollister's authority and reporting directly to Mr David Brown, with whom he had built up a good working relationship.

Between 1945 and 1949, most of Herbert Ashfield's time was taken up with developing a new crawler tractor, while still keeping his eye on tank transmissions, in particular the prototype gearboxes for the Centurion tank, which were being built in the experimental workshop. The job of running the experimental workshop fell to Stanley Mann, who became chief experimental engineer with a staff of six. Derek Marshall, who joined the experimental workshop at this time, took over field trials and was eventually put in charge of the field-test department. During this intensive period, Albert Kersey was seconded from Park Works to consult on tractor and implement designs and may have been involved in the final development stage of VAK1/C.

VAK1/C was a logical progression of the VAK1/A design and was based on groundwork prepared by Bill McCaw and Henry Merritt before they left. It incorporated the second and third phases of the VAK1 improvements, including the delayed VAK1/B modifications, the main features of which were a new gearbox, an integral power lift and a new design of mainframe.

The mainframe had to be redesigned to take a new rear-axle housing and accommodate the integral hydraulic lift. At the same time, the decision was made to split the single-piece mainframe into four separate pieces. This was done simply because only one foundry in the country was capable of casting the single-piece frame. After the war, that foundry had upped its price and David Brown did not like being held to ransom or having all its eggs in one basket.

During the VAK1/A production run, David Brown Tractors Ltd had introduced a policy of having two suppliers for every component. The idea was to minimise the risk of disruption to production schedules. If the company relied on just one supplier or foundry, then the whole of Meltham's tractor production was vulnerable if something went wrong. By splitting the frame into more manageable sections, David Brown could put the work out to other foundries, thus safeguarding its supply of castings while allowing the opportunity to search out the most competitive quotes. This explains why a number of VAK1/A tractors were supplied with a similar split-type assembly as an alternative to the single-piece mainframe. Incidentally, if supply problems ever caused a shortage of certain components, it was not unusual for the staff at Meltham to go around the works with wheelbarrows collecting up surplus parts to keep production going.

The new four-piece mainframe for the VAK1/C consisted of a centre frame (the engine lower crankcase) and a rear frame (accommodating the gearbox and hydraulic pump) and two separate footplates. The centre and rear frames were bolted and doweled together to form the lower clutch and flywheel bell-housing. The footplates bolted to the rear frame. As on the earlier models, the mainframe unit was bolted to a round front extension and machined to accommodate the rear-axle housing.

In order to accommodate the power lift, the new mainframe was slightly longer and gave the VAK1/C an increased wheelbase of 71½ in. (2½ in. longer than the VAK1/A). The rear axle was also a new casting and was offset to equalise weight distribution when ploughing, give the operator more room and improve visibility for rowcrop work. This resulted in a difference of 2 in. between the centre-line of the tractor and the centre-line of the wheels, and allowances had to be made for this when altering track widths.

The new gearbox was a six-speed unit. Designed by Henry Merritt and manufactured at Park Works, it was basically similar to the earlier four-speed unit except that the primary drive was taken to the secondary shaft through a lay-shaft. In effect, it was a three-speed gearbox with low and high ranges to double-up the speeds. The four-speed gearbox remained available as an optional extra.

The thinking behind the company's approach to gearbox design was later summarised by Herbert Ashfield in a paper on the design and development of the tractor, jointly given with Charles Birney to the Institution of British Agricultural Engineers in February 1952. According to Ashfield, 'Six speeds with a high and low reverse were adequate for all conditions, while four speeds and one reverse catered for most... Top speeds for agricultural operations should be about 7 mph, and a 15 mph transport speed could be incorporated in the six-speed gearbox. The remaining gears should then be arranged progressively between 1 and 7 mph.'

The VAK1/C's new integral power lift used a similar type of hydraulic pump as developed for the VAK2. The pump was mounted on the front end-plate of the gearbox and driven directly from the main-shaft by fixed spur-gears. The actual lift assembly was contained within the rear-axle housing. The rear axle also provided a mounting for the power take-off unit, which bolted in place of the differential cover. A combined belt-pulley and power take-off unit with two speeds was also introduced as optional equipment.

Because some of Ferguson's patents had now lapsed, the lower link-arms were provided with a choice of alternative hitch positions, allowing parallel or converging linkage arrangements. The linkage could lift up to 2,700 lb.

Prototypes of the VAK1/C were on trial by early 1946, and soon afterwards thoughts turned towards the look of the new model. A styling mock-up with all-enclosed sheet-metal appeared in mid-1946, but was dismissed as being too futuristic (and too expensive to produce). The first pre-production prototype was ready by the autumn of 1946, but this had virtually identical styling to the VAK1/A. In the end, the only noticeable change was the adoption of wider mudguards, now possible due to the easing of steel restrictions.

To coincide with the launch of the new tractor, a new paint process was introduced at Meltham with the finish baked on in ovens. The paint was still hunting pink and came from the range of Britannia Agricultural Finishes supplied by Robert Ingham Clark of London. The result was a smarter and more durable finish that gave the VAK1/C a brighter appearance.

The sales department was keen for the tractor to have a new name to underline its new features

LEFT:
A styling mock-up of the proposed VAK1/C tractor was built in 1946, but the design for the sheet-metalwork was shelved because it was thought to be too futuristic and that it would prove to be too expensive to produce.

and emphasise that it was a new model rather than a rehash of the old VAK1/A. Ford had its Major, Nuffield its Universal and International Harvester its Farmall, and so the VAK1/C became the Cropmaster. Vincent Gallagher is believed to have come up with the name, which was felt to mirror the tractor's all-round versatility.

The Cropmaster name appeared in a new script on the side of the tractor bonnet, just beneath the larger David Brown decals, which had been added at Mr David Brown's insistence because he was keen to promote the company name with the new model. The six-speed tractor was known as the Cropmaster 6 while the Cropmaster 4 had the

LEFT:
A pre-production version of the VAK1/C that appeared in late 1946 showing the eventual styling for the new tractor. It was not that different from the VAK1/A apart from the wider mudguards.

RIGHT:
A sectional VAK1/C tractor built in 1947 for exhibition and instructional purposes. Features of the new tractor included a four-piece mainframe.

RIGHT:
The VAK1/C tractor was known as the David Brown Cropmaster and looked a purposeful machine with its wider mudguards. This example has the optional wheel-weights from the auxiliaries range.

four-speed gearbox. Models with electric starting were suffixed with the letter 'S' (e.g. Cropmaster 6S).

Using David Brown's in-house alpha numeric system, tractors fitted with the power lift were designated VAK1/C/100. Tractors without hydraulics were available for those customers who did not require the power lift. These VAK1/C/200 models were known as the Cropmaster M. All VAK1/C tractor serial numbers were prefixed with a P and commenced at P10001. It may all sound complicated but David Brown's designation and numbering system was logical and made it easier for the dealer or salesman to specify the correct model variation when ordering tractors, spares or accessories.

The VAK1/C Cropmaster was launched at a gathering at Meltham plant in April 1947, and the first tractor off the line was presented to the National Farmers Union's disaster fund for the devastating floods that had followed that year's 'big freeze'. The new model was very well received by the press and customers alike. Many of its features were well ahead of its time and it became a serious contender in the burgeoning tractor market.

The new tractor was priced at £446 for the four-speed model. The six-speed gearbox cost an extra £25. The price differential was so little that most farmers opted for the Cropmaster 6 model in order to gain the benefit of the two extra gear ratios that made quite a difference when ploughing.

This lack of demand for the four-speed gearbox led to the company making the decision to delete it from the specification from 1 January 1949 and standardise on the six-speed unit, leading to a reduction in production costs. This saving was passed on to the customer and the price of the tractor was reduced by £18 on 1 March. The Cropmaster 6

LEFT:
Cropmaster VAK1/C/100 model showing the integral hydraulic lift. Alternate hitch positions were provided to allow a converging linkage arrangement. The tractor has the combined belt-pulley and power take-off attachment and an improved manual width-control hitch. Note the cable for the overload-release mechanism running from the top-link to the hand-clutch.

RIGHT:
The VAK1/C/200 model without the power lift. This tractor was marketed as the Cropmaster M. Note the extended drawbar frame.

now cost £453 while the Cropmaster 6S with electric starting was priced at £475.

The Cropmaster was a milestone model in the company's history and nearly 60,000 were made. Many of its later variations, including the diesel version, fall outside the scope of this volume and will be featured in Part Two of the work.

The VAK1/C is recognised as one of the greatest British tractors of all time, as well as one of the most famous David Brown models ever produced. To quote from the company 'The David Brown Cropmaster. True, in every way to British tradition of quality - a thoroughbred in fact.'

RIGHT:
A new paint process was introduced at Meltham to coincide with the launch of the new Cropmaster tractor. The hunting pink paint was baked on hard in ovens to give a more durable finish.

ABOVE:
The first production Cropmaster, tractor number P10001, rolls off the assembly line at the tractor's official launch, held at Meltham in April 1947.

LEFT:
The first production Cropmaster tractor built was presented to the National Farmers Union's disaster fund for the 1947 floods. Mr David Brown is seen at the wheel accompanied by the fund's representative, Mr R. N. Davies.

ABOVE:
A 1947 David Brown Cropmaster in action. The original price for the six-speed model was £471.

RIGHT:
David Brown Cropmasters on the assembly line at Meltham. Production began at around fifty tractors per week in 1947.

LEFT:
The original 1947 sales brochure for the David Brown Cropmaster tractor.

LEFT:
A David Brown Cropmaster 6S tractor working in the Vale of York in 1947. The six-speed gearbox was ideal for baling operations.

RIGHT:
An early David Brown Cropmaster with a Scottish Aviation cab. The tractor was a milestone model in the company's history and nearly 60,000 were made.

BELOW:
The David Brown range of 1948 included about a dozen implements built or approved for use with the Cropmaster tractor. The MR1 cutterbar mower featured a few minor improvements made during 1947.

Power Control

In April 1948, David Brown Tractors Ltd issued a brochure entitled 'Power Control for Better Farming'. Taking a leaf out of Harry Ferguson's book, it extolled the virtues of Cropmaster farming and described the advantages of using David Brown tractors and implements. The brochure also listed the range of equipment and machinery available, which now included around a dozen different implements.

The David Brown range was made up of equipment from two sources: implements designed and manufactured in-house at Meltham

LEFT:
The David Brown RL/2D ridger with the manual depth-control unit. The depth wheels have been set in the open diabolo arrangement to allow the ridges to be split back after planting.

BELOW:
The potato spinner remained one of the best-selling implements in the David Brown range. The latest model for the Cropmaster was known as the DP2A while the version to suit the VAK1 and VAK1/A tractors was designated DP1A.

and machinery supplied by outside companies. This latter category could then be subdivided into machines developed by David Brown dealerships and implements made by other independent manufacturers.

The machines developed by the dealers were normally introduced to satisfy the local demand for specialist equipment and to meet the needs of the local area. This included the beet lifter that had been launched in conjunction with Belton Bros & Drury, who had identified a demand for such a machine in Lincolnshire and the Fens. In a similar way, in October 1943 the south Lincolnshire dealership David Brown & Belton Ltd had introduced a front row-coverer for splitting back ridges and covering in potatoes or bulbs after planting.

Another joint project saw a steerage hoe designed by A. E. Smith manufactured on behalf of David Brown Tractors Ltd by the Midlands distributors, Bromsgrove Motors. Prototypes were tested in 1944, but the machine did not go

ABOVE:
The prototype David Brown steerage hoe on trial behind a VAK1 tractor in 1944. Manufactured as a joint project with the Midlands distributors, Bromsgrove Motors Ltd, it did not go into full production until 1949.

into production until 1949, and then sales were largely overtaken by a different type of 'flexible' steerage hoe manufactured by Belton Bros. & Drury. Middlesex dealers J. Gibbs of Bedfont made a number of introductions of matched equipment to suit the David Brown tractor, including a gang mower, park trailer and a subsoiler attachment for the single-furrow plough.

Machinery designed, manufactured and supplied for the Cropmaster tractor by outside companies formed what was known as the 'David Brown Implement and Allied Manufacturers' range. In later years, this became the 'Associated Implements' range.

The machines in this range were chosen to enhance the David Brown franchise and thereby improve the sales prospects of the tractor. The implements were usually recommended for inclusion by the company's sales personnel, dealers or even tractor owners. Occasionally, an independent manufacturer might nominate its own product for acceptance into the range. Similar approved implement schemes were operated by both Ford and Nuffield.

Once an implement was selected, it was brought to Meltham and submitted to the engineering department for test, evaluation and sometimes, modification. The manufacturers also had to meet certain provisions with regard to warranty, support and servicing arrangements before the machine could be listed as 'tested, approved and built for David Brown'. Several

RIGHT:
Newlands Works at Farsley on the outskirts of Leeds was acquired by David Brown for implement production in July 1947. The factory had been originally erected in 1941 to make shells and aircraft parts.

LEFT:
The interior of the Farsley works before David Brown moved in. Teams of engineers and maintenance staff from Meltham had the premises cleared out and machine tools installed in just over a week. The wooden box in the foreground with 'TNT' stencilled on it may account for the 'No Smoking' signs!

BELOW:
Within three weeks of David Brown occupying the Farsley plant, the factory was up to full production and rows of implements were awaiting dispatch.

tems of equipment were evaluated during 1947 and their release the following year coincided with the opening of David Brown's new implement works at Farsley.

David Brown had been preparing to release new machines of its own for some time but was restricted by the need to expand its manufacturing facilities at Meltham. Long-term plans for the extension of the existing plant were in place, but during the war the company experienced difficulty in persuading the Board of Trade to grant the necessary licences and release materials for new buildings. Government policy was to encourage firms to move into the wartime shadow factories that were now lying empty across the country rather than tie up labour and materials in erecting additional facilities. As a compromise, the Board of Trade offered David Brown redundant factories as far apart as Lincoln, Leamington Spa and Yeadon.

RIGHT:
The David Brown spike-tooth harrow was one of the first new implements to be introduced after the opening of the Farsley plant. The harrow was 16 ft wide and folded into three sections for transport. The ratchet-lever adjusted the pitch of the tines. The implement is seen at Meltham.

BELOW:
The David Brown transport or 'lift' box was another new addition to the range. It had a load capacity of 7 cwt and was designed to carry five milk churns. It is seen at Meltham being used to move oil drums.

LEFT:
The new David Brown A Series plough was the most significant of the implements to be introduced at Farsley. It had a wider cross-shaft allowing it to be used with the Cropmaster tractor's converging linkage. Note all the parts, including transport boxes, depth-wheel diabolos and plough bodies, awaiting assembly in the background.

LEFT:
The David Brown A Series plough had greater under-beam clearance and did not need the diagonal link. The two hand levers controlled the depth and front-furrow width. Initially, only a two-furrow version was available.

RIGHT:
David Brown's chief demonstrator, Nigel Palmer (left), at the official opening of the Farsley works on 8 March 1948.

BELOW:
The Paterson buckrake, manufactured by Taskers of Andover, was launched as part of David Brown's 'Allied Manufacturers' range in April 1948.

Eventually, the company earmarked premises at Farsley, near Pudsey on the outskirts of Leeds, as suitable for the satellite production of farm implements.

The factory at Farsley, known as Newlands Works, was a single-storey building covering some 80,000 sq ft. It had been erected by the Ministry of Supply in 1941 to make light shells and aircraft wings. David Brown acquired the lease in July 1947 and teams of engineers and maintenance men from Meltham moved in that autumn, clearing out the buildings and installing machine tools. With the need to get implement production flowing as soon as possible, the team worked 12 hour shifts around the clock. Amazingly, in a little over a week they had the factory ready for manufacturing to begin, albeit on a limited basis.

By early 1948, the output at Farsley trebled to the extent where implement production was running at about 250 machines per week. The production manager in charge of the satellite factory was Douglas Booth, who directed staff of some 150 workers

ABOVE:
The McConnel transportable saw bench was also included in the 'Allied Manufacturers' range. Machinery selected by David Brown for inclusion in the range had to meet strict criteria and each implement had to be tested, evaluated and approved by the engineering department at Meltham.

LEFT:
The David Brown Alley trailer outside the Aero Block in Meltham Mills Road. Made at Burnham Market in Norfolk, the 3 ton trailer had a three-way tipping facility and an all-steel body. It was added to the David Brown range in 1948.

RIGHT:
The David Brown Webb hammer-mill was made by Mancuna engineering of Manchester. The company claimed that it was the first tractor-mounted hammer-mill ever produced and it pre-dated Ferguson's Scotmec mill by two years.

mainly recruited from the Leeds area, but supplemented by a nucleus of skilled personnel seconded from the Meltham plant. The new factory took on the manufacture of the David Brown implement range, including the ploughs, cultivators, ridgers and mowers. New additions to the range included a mounted spike-tooth harrow and a transport box.

The most significant of the new introductions from Farsley was the first of the David Brown A Series ploughs with a wider cross-shaft allowing it to make use of the Cropmaster's converging linkage. The new plough no longer required the diagonal link or the manual width control and had greater under-beam clearance and new types of skimmers and discs. Initially, only a two furrow semi-digger version was available, but other models were being planned. For the time being, the standard PU series ploughs remained in production for use with the parallel linkage.

The new David Brown implements were unveiled at the official opening of the Farsley factory by the president of the National Farmers Union, James Turner, on 8 March 1948. At the same time, the first machines from the Allied Manufacturers range, notably the Robot two-row potato planter, made by Transplanters Ltd of St Albans, and the Paterson buckrake, manufactured by Taskers of Andover, were launched. More approved implements, including the McConnel transportable saw-bench and the Alley three-way tipping trailer, were launched at that year's Royal Show held at York in early July. The same show saw the appearance of the David Brown Webb hammer-mill. Made by Mancuna Engineering of

RIGHT:
Other work carried out at Farsley included the reconditioning of ex-Air Ministry tractors such as the machines shown in the photograph. Around fifteen of these were rejuvenated into 'as new' condition for sale as civilian heavy-industrial tractors every week.

Manchester, this was claimed to be the first tractor-mounted hammer-mill ever produced.

In addition to manufacturing implements, Farsley also had facilities for reconditioning the ex-Air Ministry tractors, and around ten to fifteen of these industrial machines were rebuilt there each week. A further reorganisation of David Brown's manufacturing facilities saw the assembly of both tractor engines and Aston Martin car engines transferred to Newlands Works in September 1948.

Aston Martin and Lagonda

Mr David Brown's other automotive interests are probably outside the remit of this book, but the tractor division and Aston Martin Lagonda were stablemates with common ties and often close associations. Brown's ownership of this legendary manufacturer of hand-built sports cars often had a bearing on the tractor story and the company requires at least a brief mention.

Aston Martin came to Mr David Brown's attention in 1946, when he noticed a small advertisement in *The Times*. It read: 'High-class motor business, established 25 years: £30,000: net profits last year £4,000. - Write Box V.1362, *The Times*, E.C.4.' Brown made enquiries and was surprised to find that the firm in question was Aston Martin, a well-known pre-war manufacturer of bespoke sports cars that had fallen on hard times and run short of cash.

The company had a racing pedigree that stretched back to 1914 and took its name from one of the co-founders, Lionel Martin, and the Aston Clinton hill-climb venue where he regularly competed. Although the firm produced several outstanding sporting and competition cars, it went through a succession of owners and commercial success continually eluded it . Come the end of the Second World War, the company had little to show for its endeavours other than just one prototype of a new sports saloon, the Atom, and a new 2 litre engine, both the work of Aston's chief designer, Claude Hill. With no money to put the Atom into production, the firm's owner, Gordon Sutherland, decided to put the business up for sale.

David Brown had money to invest and was keen to indulge his passion for fast cars. He visited Aston's works at Hanworth Park, Feltham in Middlesex. The works turned out to be little more than an ill-equipped and rented corrugated-iron building. Brown tried out the Atom and liked the chassis but was disappointed with Hill's four-cylinder pushrod engine. Despite this, he made an offer for the company and the deal was concluded for £20,500 in 1947.

In the meantime, Brown got wind that another prestige car firm in Middlesex, the Lagonda marque, was in financial difficulty. He knew that Lagonda had an excellent six-cylinder engine in the offing - a 2.6 litre twin overhead-cam power unit that had been designed by W. O. Bentley, who had joined the company as technical director in 1947. After being given a demonstration of the new engine, Brown put in a low bid of £52,500 for the company, which he did not expect to be accepted as he was up against some stiff competition for the marque. In the event, the other bidders pulled out and David Brown became the proud owner of both Aston Martin and Lagonda within months of each other.

The two marques were merged to form Aston Martin Lagonda Ltd in 1948. As Mr David Brown had purchased the car firms with his own personal funds, the new company stayed in private ownership, like the tractor organisation, and remained outside the David Brown & Sons group that was a public limited company. Officially, Aston Martin Lagonda was legally described as the automobile division of David Brown Tractors Ltd. Brown himself became chairman and managing director, and Allan Avison, James Whitehead and Fred Marsh were appointed to the board.

The Lagonda deal had not included its premises and so car production was set up at Feltham with David Brown taking over the lease at Hanworth Park. W. O. Bentley declined to join the new organisation and left to set up his own design consultancy, while Lagonda's Causeway Works at Staines became the home of Petter engines.

The first car to go into production at Feltham was the Lagonda 2½ litre (actually with the 2.6 litre six-cylinder engine), appearing at the end of 1947. The prototype had been designed with a French Cotal gearbox, which had naturally been replaced

ABOVE:
The Lagonda 2½ litre was the first car to go into production at Feltham following David Brown's acquisition of Aston Martin and Lagonda in 1947. The car shown is a 1948 drophead coupé version.

RIGHT:
Aston Martin Lagonda's stand at the 1948 London Motor Show showing 1949 models of the Lagonda 2½ litre saloon (left) and the Aston Martin 2 litre Sports (right). The new Aston Martin sports car became retrospectively known as the DB1, reflecting Mr David Brown's involvement in the marque.

with a David Brown unit from Park Works. Both saloon and drophead coupé versions were made, with the former model often used as a company car by senior personnel in the tractor division.

A new Aston Martin, the 2 litre Sports with Hill's pushrod engine, was launched the following year. Known retrospectively as the DB1, it featured dynamic styling, which was the work of Lagonda's chief designer, Frank Feeley, with great input from Brown himself. The car distinguished itself by winning the Spa 24-hour race in 1948, but Claude Hill resigned a few months afterwards, evidently unhappy about plans to put Bentley's Lagonda six-cylinder into his Aston chassis for the proposed DB2 model. Ironically, he went to work for Brown's old nemesis, Harry Ferguson, and was appointed chief engineer to the Ulsterman's automotive division, Harry Ferguson Research Ltd.

The DB2 was the first of the true classic Aston Martins from the David Brown stable. It was launched in April 1950 and was assembled at Feltham with the chassis and engine made at Farsley and its S430 gearbox at Park Works. The car showed great potential, particularly in the hands of the unofficial works test-driver, Mr David Brown.

TOP LEFT:
This Aston Martin DB1 Spa, exhibited at the 1948 Motor Show, was a replica of the lightweight two-seater that won the Spa 24-hours race at its first outing. The replica priced at £3,000 attracted little interest.

ABOVE
Launched in April 1950, the DB2 was the first of the true classic Aston Martins produced under David Brown's ownership. Its six-cylinder engine was made at Farsley and its four-speed gearbox was built at Park Works.

The Taskmaster

By the late 1940s, it began to look as if Mr David Brown had all the bases covered in the automotive field: Meltham was building farm tractors; Feltham, sports cars; and Park Works, an expanded range of five-speed commercial vehicle gearboxes. His post-war revival of Aston Martin and Lagonda had won him something of a reputation of a saver of lost causes, harking back to his earlier rescue of both Penistone and Meltham Mills from extinction. But he was not one to rest on his laurels and was at this time pushing the tractor division towards increasing its share of the industrial machinery market.

The New Towns Act, passed in 1946 as part of a government plan to provide much needed housing and revitalise the building industry, led to a post-war boom in construction machinery sales, which David Brown was keen to tap into. The company's industrial towing tractors had enjoyed limited success, but were heavy and expensive machines only really suited to specialist work. Production had ended following the introduction of the Cropmaster, although a few reconditioned machines continued to trickle out of Farsley until engine assembly was given priority.

A prototype semi-industrial version of the Cropmaster was built in 1947 as the first stage in developing a cheaper version of the towing tractors. It was designed with a dual-braking system to meet the needs of the Road Traffic Act

LEFT:
This prototype semi-industrial version of the Cropmaster was built in 1947. Designed to meet the requirements of the Road Traffic Act, it had dual braking, a full lighting kit, horn and rear-view mirror.

RIGHT:
The David Brown Taskmaster was launched in July 1948 as a medium-duty industrial tractor with a petrol engine, four-speed gearbox, dual-braking system and full-width mudguards. Road tyres, electric starting and a towing hitch were standard and the tractor cost £525.

with internal-expanding brakes, mounted on the rear-axle half-shafts and operated by a hand-lever, as well as foot-operated Girling brakes on the rear wheels. It was fitted with agricultural wheels but had extra weights on the rear centres.

Having identified a need for a universal machine that could be used for industrial, forestry, aircraft towing and road haulage purposes, the company decided to develop the semi-industrial into a medium-duty industrial vehicle that could also be fitted with a loader, winch or trailer-braking equipment. The new tractor was launched in July 1948 as the David Brown Taskmaster, designated VIG1/A/R.

The Taskmaster was fitted with a petrol engine, a four-speed gearbox, the dual-braking system and full-width mudguards made from steel plate. Unlike the earlier heavy-industrial tractors, it did not have a sub-frame and the mudguards were braced by the foot-plates and channel cross members, to which ballast could be added if necessary. The front fender was fitted with a central towing hitch.

Standard equipment included bolt-on steel

BELOW:
A 1948 Taskmaster with a rear-mounted belt-pulley working a grinding mill in Yorkshire. Standard equipment included a foot-throttle, rear-view mirror, speedometer and horn, and the tractor could be fitted with a loader, winch or vacuum braking systems.

ABOVE:
A Taskmaster with the optional coachbuilt cab together with a heavy-industrial tractor on David Brown's stand at the 1948 London Commercial Vehicle Show. Note the David Brown winch and Taskmaster gearbox on the left of the photograph.

disc wheels with road tyres, a six-position towing hitch, foot-throttle, electric starting, lighting with dip facility, speedometer, horn and rear-view mirror. The tractor weighed just under two tons, could tow a trailer load of up to 8 ton gross and was priced at £525.

Press correspondents and prospective customers were invited to Meltham and offered a test drive of the Taskmaster with a trailer laden with iron ingots on an exacting round trip across to the Isle of Skye Inn and back via Holmfirth and Huddersfield. A demonstration model was also fitted with a hydraulic Brayloader. Other ancillary equipment included a Boughton winch, Clayton Dewandre vacuum or Neate mechanical trailer-braking systems, and David Brown's own combined belt pulley and power take-off unit. A coachbuilt cab was also available from late 1948.

By the time the Taskmaster was launched, Fred Hollister had left the company and Herbert

LEFT:
A David Brown Taskmaster used for haulage by Brook Motors at its Huddersfield factory. The tractor was fitted with a six-position towing hitch and could handle loads of up to eight tons.

RIGHT:
The Taskmaster weighed just under two tons and it was ideal for short-haul work. Around 300 were built before the model specification changed in 1954.

Ashfield had been appointed chief engineer with full responsibility for running Meltham's engineering department. One of the first problems he faced was a request from Mr David Brown to develop the Taskmaster into a high-speed tractor with suspension, even though it was already capable of nearly 23 mph. Such machines are commonplace today, but it was forward-thinking back in the 1940s.

Ashfield's first step was to develop a sprung front-axle. This was fitted to a prototype Taskmaster, which Brown decided to take out for a test run around the moors with his chief engineer riding shotgun. The main outcome of the test was that Ashfield lost his spectacles after his boss knocked the tractor out of gear for the long downhill run into Holmfirth. No decision was made to take the project any further and the Taskmaster lost its opportunity to become the forerunner of today's JCB Fastrac.

The Taskmaster was a very popular machine that spawned a number of variations for both civilian and military use. It never really reached its full potential as a universal industrial machine and was used primarily as a short-haul tractor. Around 300 of the VIG1/A/R version were made before the specification changed in 1954.

The Century Goal

The opening of the Farsley plant had taken the pressure off Meltham in the short term. Tractor production steadily increased following the easing of restrictions on raw materials as Britain emerged from the austerity years. By early 1948, 100 Cropmasters a week were rolling off the assembly lines, allowing the company to announce that the 'century goal' had been scored.

With increased production came the problem of distribution. Tractors were dispatched from Meltham by rail from Scarr Bottom or by private haulage contractors as the company had not yet set up its own lorry fleet. Sales were helped by David Brown having a very strong and loyal dealer network in the UK. Its distributors were well organised, had good local knowledge and were passionate about hunting-pink tractors.

Meltham offered the dealers plenty of support and a sales and service training school for dealer personnel was established at the plant with lectures given in the concert room, a building that had originally been the engine house in the days when steam had powered the cotton mill.

A demonstration team was also set up in 1945. Nigel Palmer, the son of a Somerset farmer who was a committed David Brown user, was recruited in the November of that year to act as chief demonstrator and was assisted by Joe Wale. At first, the demonstrations were low-key affairs, just two or three working tractors with David Brown's public relations officer, Benet Williams, giving a running commentary through a microphone. In later years the team was strengthened and better

LEFT:
The Cropmaster was in great demand and by 1948 David Brown had pushed production up to 100 tractors per week allowing the company to announce that it had scored the 'century goal'.

equipped and its demonstrations became more extravagant.

Many of the dealers had been with the company since the Ferguson-Brown days and had remained loyal to David Brown after the Ferguson split. The sales territories were divided into six areas and some of the longest standing distributors were in Area 1, which covered Scotland, Northern Ireland and the Isle of Man. Prominent among these were Barclay, Ross & Hutchison of Aberdeen, McNeill Tractors & Equipment of Glasgow, Grassicks Garage of Blaigowrie, W. & R. Geddes of Wick in Caithness, O. D. Cars of Belfast and J. & W. Tait of Kirkwall on the Orkneys.

Dealers from other areas who originally held the Ferguson-Brown franchise included Harper & Eede of Lewes in Sussex, Drake & Fletcher of Maidstone in Kent, A. T. Oliver & Sons of Luton in Bedfordshire, Kennings of Clay Cross in Derbyshire, J. & B. Gibbs & Sons of Crewkerne in Somerset and the F. H. Burgess group in the West Country.

James E. Reed of Shiremoor, near North Shields in Northumberland, began selling Ferguson-Brown tractors in 1936. The firm also held agencies for Fordson and Case tractors before becoming exclusively David Brown distributors in the post-war years. Gibbs of Bedfont in Middlesex, owned by brothers, Murray and Sidney Gibbs, and Barton Motors of Preston, founded by Barton Townley and run by his son, Jack, were among the other leading dealers, which also included Automobile Palace of Llandrindod Wells, Ernest Doe & Sons of Ulting in Essex and Rickerby Ltd of Carlisle.

One distributor that held a special position within the David Brown retail organisation was the Lincolnshire dealership of Belton Bros & Drury. This respected firm dated back to just after the First World War, when Charles Belton and his brother, S. G. Belton, went into partnership with a Mr. H. Drury to establish an agricultural engineering and contracting business at Eastoft near Scunthorpe.

In 1936, Belton Bros & Drury became one of the first Ferguson-Brown distributors. By this time, the firm had dropped its contracting business and was concentrating on the service and maintenance of

BELOW:
A service school was established at Meltham to instruct dealer personnel and service staff in the intricacies of the David Brown tractor. Lectures were given in the works concert room, originally the engine house.

ABOVE:
Nigel Palmer with the new Cropmaster tractor at the Dublin Spring Show at Ballsbridge in May 1947. Palmer, the son of a Somerset farmer, had been recruited in 1945 to head David Brown's demonstration team.

TOP RIGHT:
David Brown's Norwich distributors put the Cropmaster tractor through its paces at a sugar beet harvester demonstration held at Raynham in Norfolk in October 1948.

RIGHT:
David Brown had a strong dealer network and its distributors were well organised. Barton Motors of Preston, originally founded in 1912 by Barton Townley, held agencies for Vauxhall, Bedford and ERF as well as the David Brown franchise for north Lancashire.

tractors and machinery. Charles Belton had built up a good reputation in the agricultural machinery trade and he later advised on the development of several David Brown implements.

In 1938, when Mr David Brown was having difficulty in finding a suitable dealer in south Lincolnshire, Belton suggested setting up a joint venture at Kirton, near Boston. He would provide the management if David Brown put up the money.

The new distribution depot at Kirton was known as David Brown & Belton Ltd and the manager was Mr H. M. Dodsworth. In 1939, the firm took delivery of the first three David Brown tractors to be dispatched from Meltham. A subsidiary depot was later opened at Fakenham in Norfolk and Tom Panton, who had joined David Brown & Belton as the office manager in 1946, succeeded Dodsworth in June 1949. Kirton was David Brown's first involvement in retail operations. Eventually, the company would own other outlets, making it unique in being a tractor manufacturer with its own retail organisation.

The export trade was not forgotten. One of the strengths of the David Brown organisation had always been its ability to generate overseas sales and contacts, and the same strategy that had been applied to gears was put into practice for tractors. On 12 August 1948, Fred Marsh and Charles Birney set sail for Rio de Janeiro to carry out an extensive survey of the South African market and meet with prospective distributors in Brazil, Argentina, Uruguay, Venezuela, Colombia and Mexico. After this they travelled on to the USA and Canada to explore the potential for tractor sales in North America.

Fred Marsh was highly skilled at encouraging overseas sales and under his direction David Brown tractors began to find their way across Europe, Scandinavia and the Middle East. Cropmasters in PKD (partly knocked-down) form shipped to New Zealand in 1947 were followed by the first sales to Australia the February of the following year. The

LEFT:
The distribution of David Brown tractors in Wales was handled by the Automobile Palace of Llandrindod Wells, run by Tom Norton. The service van is an Austin A70.

BELOW:
David Brown's Lincolnshire distributors were based at Eastoft near Scunthorpe. This highly respected agricultural engineering concern had been established in the early 1920s and became one of the very first Ferguson-Brown agents in 1936.

distribution in Australia was initially handled by British Tractors (Pty) Ltd of Sydney and Overland Ltd of Brisbane.

David Brown's first overseas tractor company was established in the Republic of Ireland in April 1948. Managed by Bill Barton, David Brown Tractors (Eire) Ltd handled sales, service and distribution across the republic from offices at Essex Quay in Dublin. It was an important time for overseas acquisitions because the following year, David Brown & Sons secured a controlling interest in Precision Equipment (Pty) Ltd of Benoni in South Africa. This new undertaking made a wide range of gears and became the largest wholly-owned David Brown manufacturing unit outside the UK.

LEFT:
The south Lincolnshire dealership of David Brown & Belton Ltd was set up in 1938 as a joint venture between David Brown Tractors Ltd and Belton Bros & Drury. The firm's extensive fleet of vehicles, consisting of (left to right) a Ford 7V truck, a Standard van, a Ford Y-Type van, an Austin van, a Riley car and an Austin car, is seen outside its Kirton premises near Boston.

ABOVE:
David Brown and Ferguson-Brown tractors and equipment outside David Brown & Belton's Kirton depot in Lincolnshire in about 1939. The company sold the first three David Brown tractors to be made at Meltham.

RIGHT:
This VAK1/A at an agricultural show in Stockholm was one of the first David Brown tractors to be exhibited overseas. The company was particularly successful in selling its products to Scandinavia.

BOTTOM:
A David Brown Cropmaster crated in PKD (partly knocked down) form for export to New Zealand in 1947. The company was keen to exploit its potential for overseas sales and actively explored markets as far afield as Australasia and the Americas.

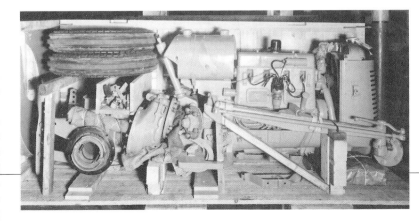

February 1949 marked the tenth anniversary of the formation of David Brown Tractors Ltd. Much had happened in a relatively short time as both tractor models and personnel had come and gone. Albert Kersey, after briefly returning to Meltham, had gone back to Park Works, where he was joined by his son, Michael, who was serving a student apprenticeship. During 1949, he was headhunted by the British Massey-Harris subsidiary to join the design team for the MH 745 tractor at the new Kilmarnock factory in Scotland.

In 1953, Kersey moved to Ford's research department at Birmingham to work on a revolutionary new hydrostatic tractor design in conjunction with the National Institute of Agricultural Engineering before being seconded to Dagenham three years later to assist John Foxwell, the executive engineer in charge of the Ford tractor division. He joined Foxwell in the United States in 1962 to work on Ford's proposed new 6X range and was responsible for the drive-line for the Ford 4000 tractor. When the 6X range was launched in 1964, Kersey returned to England but sadly died from cancer just a few years later.

Some of the people who had left Meltham, including Charles Hull, Steve Moorhouse and Fred Meadowcroft, would return to join Herbert Ashfield, who was heading up engineering with several exciting developments in the pipeline, including new agricultural and industrial tractors, crawlers, diesel engines and much more besides.

Since taking over as chief engineer after Fred Hollister left

Ashfield had moved Stanley Mann back into the drawing office and given him the responsibility for engine design and development, leaving a foreman in charge of the experimental department. Maurice Jones was also transferred to the drawing office to assist Mann on engine design.

Other draughtsmen involved in tractor design included Harry Cartwright, who had come to Meltham with the tank gearbox drawing office, and John Nichols, who had joined David Brown as an apprentice in the machine tool section in 1941 before transferring to engineering in 1947. The senior design draughtsman was George Jemmison, a quiet and retiring man, but a highly capable engineer who was responsible for the initial layout design of many of the major projects.

The marketing department was equally strong and the likes of Marsh, Whitehead and Birney had plans to widen the scope of the company's overseas influence, while Mr. David Brown had his own ideas for restructuring the organisation. Back in February 1939, David Brown Tractors Ltd had yet to produce a tractor under its own name and Meltham Mills was still making cotton thread. Within only a decade, the company had become one of the most respected names in the farm machinery business and the plant was producing some of the most advanced tractors on the market.

The next ten years looked set to be just as exciting and would see David Brown expand into one of Britain's largest exporters of agricultural machines with the broadest range of tractors and equipment in the country.

TOP LEFT:
This Swedish brochure for the Cropmaster was one of the first pieces of foreign sales literature produced for David Brown tractors.

ABOVE:
A consignment of Cropmasters are loaded at Hull docks for shipment to Lidköping in Sweden. By 1947, David Brown was enjoying export sales to Scandinavia, Europe, the Middle East, Australia and New Zealand.

LEFT:
A 1949 David Brown Cropmaster that was originally supplied by the Kirton depot. February 1949 marked the tenth anniversary of the formation of David Brown Tractors Ltd and the company looked forward to the future with a host of new models, new engines and new products in the pipeline.

Appendix 1 – David Brown Tractor Serial Numbers 1936 – 1948

VAK1 & VAG1 Agricultural 1939 - 1945

Serial No.	Date	Serial No.	Date	Serial No.	Date
1001	26/7/39	2000	2/1/41	3800	22/8/42
1050	15/1/40	2100	10/2/41	3900	24/9/42
1100	17/2/40	2200	21/3/41	4000	19/10/42
1150	19/3/40	2300	26/4/41	4100	19/11/42
1200	13/4/40	2400	25/5/41	4200	6/1/43
1250	2/5/40	2500	21/6/41	4300	12/2/43
1300	18/5/40	2600	17/7/41	4400	10/3/43
1350	1/6/40	2700	21/8/41	4500	7/4/43
1400	13/6/40	2800	13/9/41	4600	8/5/43
1450	1/7/40	2900	11/10/41	4700	8/6/43
1500	18/7/40	3000	7/11/41	4800	16/7/43
1550	9/8/40	3100	29/11/41	4900	25/8/43
1600	17/8/40	3200	24/12/41	5000	30/9/43
1650	6/9/40	3300	7/2/42	5100	30/10/43
1700	27/9/40	3400	14/3/42	5300	8/2/44
1750	30/9/40	3500	20/4/42	5600	6/7/44
1800	17/10/40	3600	27/5/42	6000	31/10/44
1900	12/11/40	3700	10/7/42	6350	6/3/45

Notes:
1. No prefix to serial numbers.
2. Final number tractor no.6350.
3. Engines use similar numbering system but are out of sequence due to industrial production.
4. VAK1/H (Narrow model) serial numbers N101 – N240.

Air Ministry Crawler 1940 – 1941

C1 – C150   June to December 1940 (Models with Hesford winch)
C151 – C551   January to October 1941 (Models with David Brown winch)

Notes: Most crawlers rebuilt as wheeled tractors with AWC prefix to serial numbers.

VIG1/100 Air Ministry (Wheeled) 1941 – 1942

AW100 – AW125 October 1941 to January 1942
AW126 – AW750 January to December 1942

Notes: Machines for civilian use have IND prefix to serial numbers.

VIG1/462 Air Ministry (Wheeled) with Turbo Transmitter 1943 – 1944
AW751- AW2000

DB4 Crawler 1941 – 1945
AC-1-1 – AC-1-110

Notes: Several military crawlers were rebuilt for agricultural use after 1946/47.

VAK1/A & VAG1/A Agricultural 1945 – 1947

Serial No.	Date	Serial No.	Date	Serial No.	Date
6351	6/3/45	8000	9/5/46	9800	13/5/47
6500	11/4/45	8500	27/8/46	9852	27/6/47
7000	21/9/45	9000	18/11/46		
7500	22/1/46	9500	31/1/47		

Notes:
1. No prefix to serial numbers.
2. First tractor no.6351.
3. Final number tractor no.9852.
4. Engines use similar numbering system but are out of sequence due to industrial production.
5. VAK1/AH (Narrow model) serial numbers N241 – N499.

VTK1 Heavy Duty (Threshing Model) 1946 – 1947
VTK1-101 – VTK1-179 (1946)
VTK1-180 – VTK1-185 (1947)

VAK1/C Cropmaster 1947 – 1948

Serial No.	Date	Serial No.	Date	Serial No.	Date
10001	21/4/47	11100	24/10/47	14500	15/7/48
10100	4/6/47	11500	10/12/47	15000	20/8/48
10200	20/6/47	12000	24/1/48	15500	23/8/48
10300	1/7/47	12500	28/2/48	16000	26/10/48
10400	18/7/47	12700	12/3/48	16500	27/11/48
10500	5/8/47	13000	7/4/48	17000	18/12/48
10900	1/10/47	14000	16/6/48	17500	10/1/49

Notes:
1. Serial numbers prefixed 'P'.
2. Prototype tractor serial no.10000.
3. First tractor serial no.10001.
4. Cropmaster model remained in production until 1953.
5. Engines in same series as VAK1 & VAK1/A models until November 1949.

VAK1/C/N Cropmaster Vineyard 1947
N10026 (5/9/47)
N10050 (31/10/47)

Notes:
1. Serial numbers prefixed 'N'.
2. Model remained in production until 1953.

VAK1/C/200 Cropmaster M 1947 – 1948
M10001 (21/4/47)
M10050 (28/2/48)

Notes:
1. Serial numbers prefixed 'M'.
2. Tractor has no hydraulic lift fitted.
3. Model remained in production until 1954.

VIG1/A/R Taskmaster 1948
R10001 (28/8/48)
R 10050 (13/10/48)

Notes:
1. Serial numbers prefixed 'R'.
2. Original series engines (as VAK1/A) used until 1949.
3. Model remained in production until specification changes in 1954.

Serial number information is given only for tractors that fall within the scope of this book (i.e. machines built up until January 1949). Later serial numbers for tractors such as the Cropmaster and Taskmaster that remained in production after this date will be given in future volumes. The lists have been compiled from David Brown Tractors Ltd communications, with additional information kindly provided by James Cochrane and Charles Clayton. It is impossible to verify the accuracy of the source material, so the figures should be regarded as a general guide only.

Appendix 2

VAK1 & VAK1/A Series Implements and Auxiliaries 1939 - 1947

1. Auxiliaries

U1 Power take-off and pulley unit
U3 Pulley for power lift attachment
U4 Rigid drawbar attachment
U6 Driver's canopy
U7 Agricultural lighting (1 front headlamp, 1 rear headlamp, battery & dynamo)
U8 Industrial lighting (2 front head/side lamps, tail lamp, horn, battery & dynamo)
U9 Electric starting equipment (battery, dynamo & starter)
U10 Agricultural lighting & starting (as U7 plus starter)
U11 Industrial lighting & starting (as U8 plus starter)
U16 Standard steel wheels – front (diameter 26"; width 5")
U17 Standard steel wheels – rear (diameter 38"; width 10")
U18 General purpose front pneumatic wheel size 4.50 x 19
U19 General purpose rear pneumatic wheel size 9 x 24
U20 Super Traction front pneumatic wheel size 6 x 19
U21 Super Traction rear pneumatic wheel size 9 x 28
U24 Rear steel wheels with retractable spuds (diameter 38"; width 9")
U25 Adjustable drawbar
U27 Front ridge-riding wheel
U28 Rear rowcrop wheel (diameter 40"; width 2")
U29 Road bands
U32 Front wheel weights
U33 Rear wheel weights
U34 Strake wheel assembly
U35 Manual width control
U36 Power lift
U36A Power lift (including lower hitch brackets)
U36B Power lift (including lower hitch brackets) with spring-loaded ball valve
U37 Overload release attachment
U38 Steel rim to take 11.25 x 24 pneumatic tyres
U39 Lower hitch bracket conversion (for use when no overload device is fitted)
U41 Lower hitch bracket conversion (for use when overload device U37 is fitted)

2. Implements

PU1 two-furrow plough 10" general purpose (adjustable 10", 11" or 12")
PU2 two-furrow plough 12" semi-digger (adjustable 10", 11" or 12")
PU3 single-furrow plough 16" deep digger
PU4 three-furrow plough 10" general purpose
PU5 two-furrow plough 10" general purpose (adjustable 8", 9" or 10")
U33 Conversion set to convert PU1 to three furrow
U34 Conversion set to convert PU5 to three furrow
RL1D Three-row ridger without depth control
RL2D Three-row ridger with manual depth control
RL3D Three-row ridger with dual wheel depth control
U12A Manual depth control for ridger or cultivator
U14 Dual wheel depth control for ridger or cultivator
U42 Ridger fin conversion unit
CS1A Spring tine dual purpose cultivator without depth control
CS2A Spring tine dual purpose cultivator with manual depth control
CS3A Spring tine dual purpose cultivator with dual wheel depth control
CR3 Rowcrop cultivator with dual wheel depth control
CR5 Rowcrop cultivator with manual depth control
DP2 Potato spinner
BR1 Beet lifter
MR1 Cutterbar mower
SU1 Unit principle sprayer

Bibliography

Journals & Periodicals

British Engineering
British Power Farmer
Classic Car Weekly
Commercial Motor
Contact (David Brown Publication)
Daily Telegraph
David Brown Tractor News
The Engineer
Engineering
Farm Implement & Machinery Review
Farm Mechanisation
Farmers Weekly
Huddersfield Daily Examiner
The Independent
Journal of Automotive Engineering
Journal of the Tractor Users Association
Marketplace
Newsletter (David Brown Publication)
Power Farming
Railway Gazette
Sunday Chronicle
Telegraph & Argus
The Times
Tractor News (David Brown Tractor Club)

Books

Ashfield, H. E., **My Career as an Engineer**
(David Brown Tractor Club, 2000)

Ashfield, H. E., **More on My Career as an Engineer**
(David Brown Tractor Club, 2003)

Batchelor, John H., and Macksey, Kenneth, **Tank**
(Macdonald & Jane's, 1970)

Bowler, Michael, **Aston Martin – The Legend**
(Parragon Publishing, 1997)

Broatch, Stuart, and Townsin, Alan, **Bedford 1923-1950**
(Venture Publications, 1995)

Calder, Angus, **The People's War**
(Jonathan Cape, 1969)

Craven, Leonard, **Season of Growth**
(unpublished manuscript, 1977)

Donneley, Desmond, **David Brown's - The Story of a Family Business** (Collins, 1960)

Earnshaw, Alan, **David Brown Tractors 1936 - 1964**
(Trans-Pennine Publishing, 1997)

Fraser, Colin, **Harry Ferguson - Inventor & Pioneer**
(John Murray, 1972, Old Pond, 1998)

Gibbard, Stuart, **The Ferguson Tractor Story**
(Old Pond, 2000)

Gibbard, Stuart, **Ford Tractor Conversions**
(Farming Press, 1995, Old Pond, 2003)

Harvey, Chris, **Aston Martin and Lagonda**
(Oxford Illustrated Press, 1979)

Hughes, Joseph, **A History of the Township of Meltham** (J. Crossley, 1866)

Merritt, H. E., **Gears** (Sir Isaac Pitman & Son, 1942)

Walker, Allen, **Inside Story**
(David Brown Tractors Ltd., 1978)

Other Publications

Agricultural Machinery - A Report on the Industry
(Political & Economic Planning, 1949)

Data Book of RAF Vehicles
(Ministry of Supply / Air Ministry, various issues)

The Brown Family of Yorkshire
(Debrett, 1982)

The Development of Toothed Gearing
(The David Brown Organisation, undated)

The First Ten Years
(The David Brown Organisation, 1948)

Wartime Activities & Products
(David Brown Tractors Ltd., undated)

Worm Gearing
(David Brown & Sons (Huddersfield) Ltd., 1920)

Technical Papers

Ashfield, H. E., and Birney, J. C. R.,
The Design and Development of a Tractor
(Institution of British Agricultural Engineers, 1952)

Merritt, H. E., **The Evolution of a Tank Transmission**
(Institution of Mechanical Engineers, 1945)

Various David Brown Corporation, David Brown & Sons (Huddersfield) Ltd and David Brown Tractors Ltd journals, sales and service publications.

Index

Roman numbers refer to text, bold to illustrations

A

A. C. Cars, 111
A. T. Oliver, 201
Admiralty, The, 80, 135, 137
Aircraft
    Avro Lancaster, 83, 134
    Avro Manchester, 128
    Boulton Paul Defiant, 83
    De Havilland Dove, 150
    De Havilland Mosquito, 83, **84**
    Fairey Battle, 81
    Halifax, 130
    Handley Page Halifax, 83, 128
    Handley Page Hampden, **138**
    Hawker Hurricane, 82-3, **83**
    Lockheed Ventura, **134**
    Short Stirling, 83, 128, **133**
    Supermarine Spitfire, 60, 81-5, **82**
    Vickers Vimy, 123
Air Ministry, 112, 121, 123-6, 128-9, 131, 134-5, 137
Allen, Tommy, 133
Alley trailer, **193,** 194
Allis Chalmers, 35, 139
    All-Crop 60 combine, **116**
    Model U tractor, 61
Anderson, Andy, 132
Annatt, Bob, 43, 48, **48,** 53
Arlborough-Smith, Captain Arthur, 43, 48, 97, 112 152
Armitage, Arthur Carlow, 78
Armitage, Fred, 40, **48,** 96
Ashfield, Herbert, 87-8, 90, 92, 111, **112,** 135, 137-8, 153-6, **153, 159,** 166, **171,** 177-8, 200, 204
Associated Equipment Company (AEC), 16
    Monarch, **115**
    Y-type truck, **17**
Aston Clinton hill climb, 195
Aston Martin Lagonda Ltd, 195
    Aston Martin cars, 195, 197
        2 litre Sports, (DB1), 196, **196**
        Atom, 195
        DB1 Spar, **197**
        DB2, 196, **197**
    Feltham works, 195-7
    Lagonda cars, 195, 197
        Lagonda 2 _ litre, **56, 195, 196**
Atkinson lorry, **161**
Austin 10 van, **47, 97,**
Automobile Palace, 201, **203**
Automobiles Gregoire, 111
Automotive Products Company, 98
Avison, Allan, 28, 81, 150, 195

B

Badsworth hunt, 27, 150
Barclay, Ross & Hutchison, 105, 201
Barton, Bill, 152, 203
Barton, Jack, 201
Barton Motors, 201, **202**
Barton, Townley, 201
Beaverbrook, Lord, 83, 90, 110, 122-3, 127
Belton Bros & Drury, **119,** 120, **163,** 164, 186, 201, **203**
    Beet lifter (see David Brown implements)
    Steerage hoe, 188
Belton, Charles, 201
Belton, S. G., 201
Bentley Motors, 20, 27
Bentley, Walter Owen, 19,195
Bertram Mills Circus, 172, **172**
Birney, Charles, 152, 178, 202, 205
Blackwell, Arthur, 52, 88, 96, 152
Booth, Douglas, 192
Booth, George, 137
Booth, Jack, 88
Bostock, F. J., 15
Brabazon, Lord, 110, 123, 150
Bristol Aircraft, 111, 172
British Tractors (PTY), Ltd, 105, 203
Broadbent, Thomas, 12, 111
Brockhouse Engineering, 132
Brook, Charles, JP, 78
Brook, Jonas, 76
Brook Motors, 172, **199**
Brook, William, 76
Bromsgrove Motors, 187
    Steerage hoe, 187-8, **188**
Brown & Broadbent, 12
Brown, Angela, 27
Brown, Carrie (nee Brook), 19, **19,** 25
Brown, David, (founder, b. 1843), 12-13, **12**
Brown, David (later Sir, b. 1904), **19,** 22-3, 25-8, **26, 27,** 3 34, 37, 44, 46, 48, **48,** 54-6, 61-2, 64-8, 71-2, 74-5, 77, 78, 83, 91, 109-11, **109,** 122, 129, 138, 146,150, **151, 152,** 153, 164-6, 177, 179, **183,** 195, 197, 200, 205
Brown, David, (junior, b.1927), 27
Brown, Ernest, 12-13
Brown, Francis Edwin (Frank), 12, 14-15, 23, 25, **25,** 27, 37-8, 65, 109
Brown, John, 12
Brown, Percy, 13, 15, 22, 25, 37
Bryce Ltd, 111
Burgess, Frederick, 17, 19-20, 26

C

C.J.Fitzpatrick, 81
Campbell, Donald, 87
Campbell, Sir Malcolm, 87
Carr, Albert, 84
Cartwright, Harry, 205
Case tractors, 201
    Model L, 140
Caterpillar, 139-42
    D4, 139-42, **140, 141,** 145
    D6, 139
    D7, 139
    D8, 139
Caudwell, Arthur, 138
Chamberlain, Neville, 75
Chambers, John, 35, 44, **44,** 53
Churchill, Winston, 60, 90, 123, 144
Citroen, 111-12,
Cleare, Vernon, 86-7
Clement Talbot (later Sunbeam Talbot), 111
Cleveland Cletrac, 139
Coats, Stanley, 88
Cobb, John, 87
Cockshutt plough, 34
Commer Cars, 172
Coventry Climax Engines Ltd, 41-5, **41,** 62
Craddock, T. Hammond, 152
Craven Wagon & Carriage Works, 37
Craven, Len, 25, 44, 53, 79, 97, 118, 152
Crichton, Jack, 166
Cripps, Sir Stafford, 152, **152**
Culpin, Claude, 165

D

Daimler, 111
    Dingo scout car, **85**
David Brown & Belton Ltd, 186, 202, **203, 204**
    Front row-coverer, 186
David Brown & Company, 12
    Chapel Street works, 12
David Brown & Sons, 13, 109
    East Parade, 12-13, **13**
David Brown & Sons (Huddersfield) Ltd, **14, 15,** 14-17, 20, 22-3, 25, 27-8, **30,** 30-1, 37-8, 60, **60,** 64, 81, 87, 109-10, 150
    Cars, 17-19
        Dodson, **18,** 19
        Sava, 19
        Valveless, **17, 18,** 18-19, **19,** 26
        Park Works, **10-11,** 13, **13, 14, 15,** 16-20, **21, 22,** 22, 25-30, 36-7, 40, 42, 44, 48-53, **49, 50, 51, 52,** 54, 60-1, 64, 66, 68, 70, 75, 78-9, **80,** 81-3, **85,** 86-9, 92-3, 96, 108, 110, 112-13, 123, 177-8, 196-7, 204
    Patent Thread, 15-16, **16**
    Penistone Foundry, 23-5, **24,** 37, 44, 48, **61,** 78, 81, **81,** 85, 150, 197
    Radicon gear unit, **29,** 30

David Brown Corporation, 28
David Brown Gear Industries Ltd, 92
David Brown Gears (London) Ltd, 81
David Brown implements, 162, 186, 101, 104,194, 202
    Beet lifter 164, **163,**186
    Cultivators, **104,** 162, 194
    Hammer-mill (see Mancuna Engineering)
    Mower, 162-3, **162, 186,** 194
    Ploughs, **103,** 162, 167, 194
        A Series, **191,** 194
        PU Series, 194
        Single furrow, 104
        Three furrow, 104, 119
        Trailing (PW1), 101, **101,** 104
        Two furrow, **102,** 104, **104**
    Potato spinners, 119-20, **119,** 162, **187**
    Ridgers, 105, **105, 106,** 119, **187,** 194
    Spike tooth harrow, **190,** 194
    Sprayer, SU1, 163, **163**
    Steerage hoe, **188**
    Tipping trailer (see Alley)
    Transport box, **190,** 194
David Brown Implement & Allied Manufacturers, 188, 194
David Brown tractors
    Air Ministry Tractors, **83, 84, 109,** 122-38, **129, 130, 131, 132, 133, 137, 138,** 150, 169, 172-5, **175, 194,** 195
    VIG1/100, 130-1, 133-4, **134, 135, 150,** 169-70, **170**
    VIG1/462, 134-5, **136, 137,** 138, 172
    Crawlers, 124-7, **124, 125, 126, 127, 128,** 131-2, **132, 135,** 139, 141, **141**
    Cropmaster (see VAK1/C)
    DB4 crawler, **94-5,** 139-47, **139, 140, 142, 143, 144, 145, 146, 147,** 175
    Industrial Tractors, 174
        Prototype semi-industrial, **123,** 124
        Cropmaster semi-industrial, 197-8, **197**
        IND series, 131, **135,** 138, 169-70, **172, 173, 174, 175, 199**
        VAK1/A, medium industrial, 169, **169**
        VIG1/100, (see Air Ministry tractors)
        Taskmaster (VIGI/A/R), 197-200, **198, 199, 200**
    Prototype (first), **62, 64, 65,** 67-72, **67, 69, 70, 71, 72**
    VAG1, 113, 123, **124**
    VAK1, **59, 60,** 61, **65,** 73, **73,** 74, **74,** 75, **97, 98,** 99-101, **99, 100, 101, 102, 103,** 106-8, **106, 107, 108,** 113-17, **113, 114, 115, 116, 117,** 120-1, **121, 122,** 153-60, **153,** 164-5, 177, **188**
    VAK1/A, **153, 154,** 155-62, **155, 156, 157, 158, 159, 160, 161, 162,** 164, 169, **171,** 172, 176-9, **204**
    VAK1/B, 176-7, **176, 177**
    VAK1/C (Cropmaster), **148-9,** 161, 176-8, **177, 179, 180,** 181-2, **181, 182, 183, 184, 185,** 186, **186,** 188, **190, 192, 193,** 194, 197, **201,** 202, **202, 204, 205**
    VAK2, 154,164, **164,** 166, 176, 178
    VTK1, (threshing model), 170, **170, 171, 172**

David Brown Tractors Ltd, 38, 39, 48, 57, 66, 73, 80, 90, 92, 99, 102, 108, 110-13, 140, 146, 150-2, 161, 165, 167, 174, 176-7, 186, 192, 203-5
   Aero Block, **59**, 80, **81**, 82-3, **84**, **85**, **97**, **128**
   Aero-gears, 81-5, **82**
   Meltham Mills factory, 57, **59**, 75-6, **76**, **77**, 78-80, **78**, **79**, **80**, 82-4, 88-9, **88**, **91**, 94, **96**, 98, **98**, **99**, 103, 106, 108-9, **110**, 111-12, **114**, **117**, **118**, 119, 121, **122**, **130**, 131, **135**, 136-7, 140-2, **143**, **144**, 145, 146, 150, **151**, 152-3, **152**, **161**, **162**, 163-4, 167-9, 173-4, 176, 178, 181, **182**, **183**, **184**, 186, 188-9, 194, 197, 200, **201**, 204-5
   Newlands Works, Farsley **175**, **176**, **188**, 189, **189**, 192, **192**, 195, 197, 200
   Scarr Bottom, 76, **77**, 84, 91, **174**, 200
David Brown Tractors (Eire) Ltd, 203
Davies, R. N., **183**
Delaunay-Belleville, 111
Derby Council, 172, **173**
Doe & Sons, 201
Dodson, Mr, 18
Dodsworth, H. M., 202
Drake & Fletcher, 201
Drury, H., 201
Durker Roods, 68, 78, **78**, 91
Dyson, Nathaniel, 76

E

Eros tractor, 35
Eyston, Captain George, 87

F

F. H. Burgess, 201
Fairfield Shipbuilding Company, 20
Feeley, Frank, 196
Ferguson, Harry, 34, **34**, 35, 36, 37, **38**, 39, **39**, 44, 47, 48, 53-6, 66-7, 74, 75, 101-4, 109, 178, 186, 196
Ferguson-Brown Ltd, 48, 52, 56, 57
Ferguson implements, **32-3**, 42, **51**, 104
   'Belfast' plough, 35, 36
   Type B, two-furrow plough, 42, **42**
   Type C, spring tine cultivator, 42-3, **42**
   Type D, three row ridger, 43, **43**
   Type E, spring tine cultivator, 43, **43**
   Single-furrow plough, 43
Ferguson tractors
   'Black Tractor', 35-7, **35**, **36**, 39, 41
   Type A, **32-3**, 37- 42, **38**, **39**, **40**, **41**, **42**, 44-8, **44**, **45**, **46**, **47**, **50**, 53-4, **53**, **54**, 55, **55**, **56**, 57, **57**, 61-6, 68, 74, **74**, 107, 201
   Ferguson-Brown tractor (see Ferguson Type A)
Finnmark, 160-1
Firth, Daisy, 26
Fleet Air Arm, 125, 135
Fletcher, Rivers, 28
Ford, Henry, 35, 55-6

Ford Motor Company, 56, 94, 108, 188
   Ford tractors
      4000, 204
      6X range, 204
      Fordson tractors, 61, 65, 94, 107, 108, 128, 201
         Major, 179
Foxwell, John, 204
France, Edgar, 88

G

Gallagher, Vincent, 110, 152, 179
General Motors Corporation, 62, 63
Gibbs of Bedfont, 188, 201
   Gang mower, 188
   Park trailer, 188
   Subsoiler, 188
Gibbs, Murray, 201
Gibbs, Sidney, 201
Gilford Motor Company, 46
Girling Ltd, 98
Gladwell, Arthur, 54
Gladwell & Kell Ltd, 54, 64
Glasgow Corporation, 172, **173**
Gower, Hubert, 166
Grassicks Garage, 201
Greer, Archie, 35, 44

H

H. O. Serck Ltd, 98
Halifax Corporation, 175, **175**, **176**
Hammersley, S. S., 110
Handley Page, 172
Harper & Eede, 201
Harrison, Bill, 43, 48, **48**, 67-8, 103
Harry Ferguson Ltd, 39, 43
Harry Ferguson Research Ltd, 196
Hartree, Professor, 87
Hercules Corporation, 36, 41
Hesford, C. M., 125
Hill, Claude, 195-6
Hill, Walter, 43, 48, **48**, 96, 102, 108
Holland, Jack, 146, **146**, 175
Hollister, Fred, 165-7,176,199, 204
Huddersfield Corporation, 15, 146, **174**
   Trolley-bus, **31**
Hudson, Robert, Spear, 123
Hull, Charles, 66,103-4, 108, 111-13, **112**, **113**, 126, **153**, 155-6, 164, 177, **177**, 204
Humber, 19

I

International Harvester Company, 168
   Farmall tractor, 179
Institution of British Agricultural Engineers, 178
Issigonis, Alec, 167

J

J. & B. Gibbs, 201
J. & P. Coats Ltd, 77, 78
J. & W. Tait, 201
J. I. Thorneycroft, 20
Jackson, P. R. Ltd, 37
James E. Reed, 201
James, Sidney, 137
JCB Fastrac, 200
Jemmison, George, **112,** 205
Jenkins, Bert, 109, 116
John Deere tractors
    Model B, 61
John Fowler of Leeds, 139
John Samuel White, 20
Jonas Brook & Bros. Ltd, 76, 77
Jones, Maurice, 113, 205
Joyce, William, (Lord Haw-Haw), 82

K

Karrier Motors, 39, 43, 52, 96
    Cob semi-trailer unit, **32-3**
Keighley Gear Company, 22, **23**
Kennings, 201
Kenyon, Ernest, 53, 67-8
Kersey, Albert, 46-7, **48,** 55, 61-8, 74, 86, 88, 96, 101-3, 105, 107-8, 110-12, **112,** 124, 155, 164, 177, 204
Kersey, Michael, 204
Ketchell, Bill, 53
Knox, Trevor, 43

L

Lazenby, Tom, 152
Leyland Motors, 112
London General Omnibus Company, 16
Lucas, Ralph, 17-18

M

McCaw, Alan William, 65, 66, 96, 105-6, 111, 130, 132-3, 135, 177
McConnel saw-bench, **193,** 194
McHardy, D. N., 152
McNeill Tractors & Equipment, 120, 201
Mancuna Engineering, 194-5
    Webb hammer-mill, 194, **194**
Mann, Stanley, 112-13, **112,** 177, 204-95
Markland, Stanley, 112
Marsh, Fred, 87-8, 90, 151-2, **151,** 195, 202, 205
Marshall, Cyril, 43, 48, 67-8
Marshall Derek, 177
Marshall, Ronnie, 152
Martin, Lionel, 195
Massey Harris, 167, 204
    Tractors,
        Pacemaker, 61
        745, 204

Matthews, Mary Jane, 12
Maybach, Dr., 87
Maybach Gears Ltd, 87
Mays, Raymond, 26-8
Meadows, 26, 62, 64
Meadows, Henry, 64
Meadowcroft, Connie, 83
Meadowcroft, Fred, 43, 45, 47, 48, **48,** 53, 62-3, 68, 70, 83, 101, 104-5, 111, **113,** 120, 126, 154, 159, 162, 167-9, **167,** 204
Meltham Hall, 76, 77, **77**
Merritt, Dr Henry, 28, 85-7, 89-90, 105-6, 109, **109,** 111-13, **112,** 121, 130, 135, 153-55, **153,** 164-7, 169, 177-8
M'Ewan, Lieutenant Colonel Ewan G., 87
Ministry of Agriculture, 116, 139, 146
Ministry of Aircraft Production, 80, 83
Ministry of Home Security, 80
Ministry of Supply, 75, 80, 98, 111, 123, 133, 137, 139, 143, 173, 192
Moes, Emil, 96
Moorhouse, Steve, 111, **112,** 164, 166-7, 204
Morris Motors, 35, 96, 108-9, 127, 167
Muir-Hill, 135
Muir Machine Tool Company, 150

N

National Farmers Union, 151, 194
National Institute of Agricultural Engineering, 113
Navy & Vickers Gear Research Association, 150
Nichols, Frank, 205
Norris, Henty & Gardner, 112-13
Nuffield Mechanisations & Aero Ltd, 87, 91, 165
Nuffield Organisation, 91, 121, 164-7, 169, 176, 188
    Universal tractor, 164-8, **166, 167, 168,** 179
Nuffield, Viscount, (formerly William Morris), 165

O

O. D. Cars, 201
Overtime tractor, 34

P

P. R. Jackson, 22, **23**
Palmer, Nigel, **116, 148-9, 172, 192,** 200, **202**
Panton, Tom, 202
Patrick & Wilkinson, 163
Pilkington, Harry, 40, 52, 92, 96
Precision Equipment (PTY) Ltd, 203

R

Ramsden, Arthur, Maxwell, 109-10, 166
Ransomes & Rapier, 35
Ransomes, Sims & Jefferies, 165
Ratcliffe, Cyril, 48
Reekie, George, 118
Renault, 19, 111
Ricardo, Harry, 26, 64, 112, 140

Richardson Westgarth, 20
Rickerby Ltd, 201
Roadless Traction Ltd, 124-5, 139
    Rushton Roadless, 124
    Case Roadless, 139
Robert Ingham Clark, 178
Roesch, George, 111, 155
Rolls-Royce, 60, 62, 81, 118, 172
    Merlin, 81-3
    Peregrine, 83
    Vulture, 83
Roosevelt, President, 123
Rotary Hoes, 135
Rover Company, 35
Royal Agricultural Society of England, 108
Royal Air Force, (RAF), 83, 123, 125, 127-8, 130-2
Royal Engineers, 125, 139-40
Royal Navy submarine, **20**

S
Sands, Willie, 34, 35, 44
Scammel Mechanical Horse, 49, **98**
Schneider, Ernst, 87, 92
Scott, H. P, 108, 151
Scott, Reg, 108
Seagrave, Sir Henry, 87
Sidney, Edgar, 111,
Simms Motor Units, 112, 177
Société Anversoise pour Fabrication des Voitures
    Automobiles, 19
Smith, A. E., 111, 119, 187
Stanhay Ltd, 160
Steels Engineering Products, 135
Steeds, W, 87
Stirling, James, 96, 112
Street, A. G., 152, **152**
Sugar Beet Education & Research Committee, 164
Sutherland, Gordon, 195
Sykes, Arthur, 28, 30, 40, 44, 48, 61, 86-7

T
T. Baker & Sons (Compton) Ltd, 119
Tanks, 87
    Avenger, 90
    Centaur, 90
    Centurion, 92, 177
    Challenger, 90, 92
    Challenger 2, 93, **93**
    Char B, 86
    Chieftain, 92
    Churchill, **88**, 89-90, **89**
    Comet, 90, **92**
    Conqueror, 92
    Covenantor, 87, 89
    Cromwell, 90
    Crusader, 87
    Matilda, **85**
    Tortoise tank destroyer, 90
    Valentine, **85**
    Vickers A-6, 86
Tank gearboxes, 31, 87-93, **91**, 177
    Merritt-Brown, 85-90, **86, 88,** 92, **92**
    Merritt-Maybach, 87
    Meadows-Wilson, 87, 92
    Nuffield-Wilson, 87
Taskers of Andover, 194
    Paterson buckrake, **192**, 194
Taub, Alex, 62, 63, 90
Thomas, Sir Miles, 91, 165-6
Thompson, Harold, **48**, 52
Thornycroft army lorries, 20, **20**
Timken Detroit Axle Company, 15
Tractor Spares Ltd, 142, 146
Tractor Traders Ltd, 142
Transplanters Ltd, 194
    Robot two-row potato planter, 194
Tuplin, William, 28, 89
Turner, James, 194

U
United Thread Mills, 77-9

V
Vauxhall Motors, 62, 63, **89**, 90, 96, 164
    Bedford, **62**, 63, 64, **97**
    Model 10 car, 63
    Vauxhall Villiers racing car, 26-7, **26,** 62
Vickers Armstrong, 172
Villiers, Charles Amherst, 26-7

W
W. & R. Geddes, 201
W. E. Bray, 139
W. H. Dorman & Co. Ltd, 140-1, **140,** 145-6
Wale, Joe, 200
Walker, Herbert, 96
Wallace & Steevens Advance steamroller, **28,** 29
Wallis, Bill, 97, 118
Wallsend Slipway, 20
War Agricultural Executive Committee, 161, 170
War Office, 28, 116, 137, 139
Weight, Charles, 142, 146
Westland Aircraft, 96
Whitehead, James, 150-2, **151,** 195, 205
Willey, Harold, 43, 53
Williams, Benet, 200
Williams, Tom, 151-2
Wilson, Major Walter G., 86-7
Wisner, Professor, 111-12, 177

Y
Yarrow, 20

Z
Zeppelin, Count, 87

The David Brown Tractor Club

The David Brown Tractor Club Ltd would like to thank all members and past employees of David Brown Tractors Ltd for their assistance and contributions to Stuart Gibbard in the production of this book.

For information on David Brown Tractors or the David Brown Corporation, please contact:

David Brown Tractor Club Ltd
PO Box 990
Holmfirth
HD9 1YH
United Kingdom

Fax (0) 1484 690454
Website www.davidbrowntractorclub.com

Books and Videos from

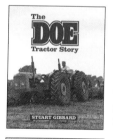

The Doe Tractor Story
Stuart Gibbard

With its two engines, 4-wheel drive and 90-degree articulation, the Doe Triple D was one of the most unorthodox of tractors. This is a detailed account of Triple D and the associated tractors and machinery from Ernest Doe & Sons, written with full access to the Doe archives and illustrated with many previously unpublished photographs.
Hardback book, 120 pages inc. 190 photographs. ISBN 1-903366-17-8. **$17.95**

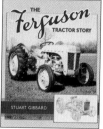

The Ferguson Tractor Story
Stuart Gibbard

This highly illustrated book covers the full history of Harry Ferguson's tractor developments from the Belfast plough, through the Ferguson-Brown and Ford Ferguson to the massive production run of the TE-20 at Banner Lane. The story concludes with the days of Massey-Harris-Ferguson and the FE35 tractor. Overseas production, prototypes, variants and industrials are fully covered, as is the use of Ferguson equipment around the world.
Hardback Book, 168 pages inc. 250 photographs. ISBN 1-903366-08-9. **$19.95**

Ferguson on the Farm Part One
Harold Beer and Stuart Gibbard

Specially filmed on Harold Beer's family farm in North Devon to show a wide range of Ferguson implements at work. Part One focuses on the potato crop and grass and includes many associated activities. Details of the implements are shown and the full commentary is scripted by Stuart Gibbard.
VHS video, 60 minutes. ISBN 1-903366-19-4 **$15.95**

Ferguson on the Farm Part Two
Harold Beer & Stuart Gibbard

This video shows a further 20-25 Ferguson implements at work on Harold Beer's family farm. Among the items included are ploughs (1, 2 and 3-furrow), trailed and mounted disc harrows, sub-soiler with spade lugs, Ferguson crane, dump skip, winch, cordwood saw and multi-purpose corn drill. There is a complete sequence showing the hammer mill and another using a muck-spreader, both featuring different trailers and loaders. A rare row-crop thinner is shown at work on sugar beet.
VHS video, 65 minutes. ISBN 1-903366-22-4 **$15.95**

Fifty Years of Farm Tractors
Brian Bell

Dealing with over a hundred companies, in an A-Z format, Brian Bell describes the models and innovations that each contributed to post-war tractor development. With over 300 illustrations, the book is a unique guide to the wide range of machines to be found on Britain's farms.
Hardback book, 256 pages inc. 320 illustrations. ISBN 0-85236-525-X **$19.95**

Ransomes and their Tractor Share Ploughs
Anthony Clare

With original research in the Ransomes archives, Anthony Clare classifies and shows the ploughs produced by the company. Deals with the horse ploughs adapted for tractors, mounted plough development, Ford-Ransomes, the links with Dowdeswells and includes analyses of identification codes and the TS classification.

About the Author

A successful author and journalist specialising in tractors and machinery, Stuart Gibbard comes from a farming background near Spalding in Lincolnshire. He has developed his interest in collecting early tractor literature into a mail-order business which is run by his wife Sue, and he is also one of the founders of the annual Spalding model-tractor and literature show.

Stuart's first book, published to much acclaim in 1994, was *Tractors at Work*, a remarkable collection of rare and archive photographs spanning ninety years of tractor development from Dan Albone's 1904 Ivel. Stuart has since produced another eleven tractor books and six videos, including prize-winning titles on Ford and Ferguson. He is the founding editor of *Old Tractor* magazine.

Devoting much of his time to historical research, Stuart has talked to many of the men who played their part in creating the machines portrayed in his books and videos. This first-hand knowledge has enabled him to give a fascinating insight into the world of agricultural engineering and tractor development.

Publications

BOOKS

Tractors at Work: a pictorial review 1904-94 (1994).

Ford Tractor Conversions: the story of County, Doe, Chaseside, Northrop, Muir-Hill, Matbro and Bray (1995, 2003)

Tractors at Work: a pictorial review Volume Two (1995).

Roadless: the story of Roadless Traction from tracks to tractors (1996).

Change on the Land: a hundred years of mechanised farming (1997).

County: a pictorial review (1997).

The Ford Tractor Story Part One: Dearborn to Dagenham 1917 to 1964 (1998).

The Ford Tractor Story Part Two: Basildon to New Holland 1964 to 1999 (1999).

The Ferguson Tractor Story (2000).

Tractors in Britain (2001)

The Doe Tractor Story (2001)

VIDEOS

Roadless Tractors (1996).

County Tractors (1997).

Ferguson Tractors (1998).

Ferguson on the Farm Part One (2001).

Ferguson on the Farm Part Two (2002)

Giants of the Field (2003)